Crop Science and Technology

Crop Science and Technology

Edited by
Cassius Foster

Larsen & Keller
www.larsen-keller.com

Crop Science and Technology
Edited by Cassius Foster
ISBN: 978-1-63549-018-3 (Hardback)

© 2017 Larsen & Keller

Larsen & Keller

Published by Larsen and Keller Education,
5 Penn Plaza,
19th Floor,
New York, NY 10001, USA

Cataloging-in-Publication Data

Crop science and technology / edited by Cassius Foster.
    p. cm.
Includes bibliographical references and index.
ISBN 978-1-63549-018-3
1. Crop science. 2. Agriculture. 3. Agricultural innovations. I. Foster, Cassius.
SB91 .C76 2017
633--dc23

For more information regarding Larsen and Keller Education and its products, please visit the publisher's website www.larsen-keller.com

Table of Contents

Preface

Crop science is an integral part of agricultural practice and it studies the fundamental practices of agriculture such as agronomy, crop production, soil, and pest management. It focuses on local and global condition related to climate, soil and the specificity of crops. The field is a multidisciplinary in its approach. This book is compiled in such a manner, that it will provide in-depth knowledge about the theory and practice of crop science. Most of the topics introduced in this text cover new techniques and applications of this field. Coherent flow of topics, student-friendly language and extensive use of examples make this book and invaluable source of knowledge.

A foreword of all chapters of the book is provided below:

Chapter 1 - Agricultural science can be defined as the application of agricultural techniques in compliance with sustainable development practices. Agricultural science is deeply integrated with global food production and consumption. The practice of agricultural productivity and pest management is a key component of this discipline. This chapter is an overview of the subject matter incorporating all the major aspects of agricultural science; **Chapter 2** - Agricultural practices vary according to location as well as the demand for food products in an area. This is why there are a variety of agricultural practices that exist and are utilised alongside each other. The topics discussed in the chapter are of great importance to broaden the existing knowledge on agricultural science; **Chapter 3** - Agricultural practices related to the growth and harvesting of crops is discussed in this chapter. Careful attention is given to techniques related to weed control and plant breeding. The chapter on agricultural cycle offers an insightful focus, keeping in mind the complex subject matter; **Chapter 4** - Tests and observations required by agricultural science are performed with the help of its integrated branches of study. Some of these fields are agrophysics and agricultural economics. This chapter elucidates the crucial theories and principles of agricultural science; **Chapter 5** - Growing concerns about irreparable environmental damage and pesticide related diseases in humans and animals have highlighted the importance of alternative agricultural systems. Discussed in this chapter are practices like organic farming, rainfed agriculture and cash crops. This chapter discusses the methods of alternative agriculture in a critical manner providing key analysis to the subject matter; **Chapter 6** - Intensive farming practices can sap the essential nutrients that are naturally found in the soil. These nutrients are essential for the growth and developments of plants. In this chapter, techniques such as hydroponics that deliver energy to plants and nutrition restoration are discussed; **Chapter 7** - Recent advances in crop production techniques have led to bumper crops and high-yielding variety seeds. Such technologies enable farmers to gather greater output of crops during harvest while spending less time and energy. Tools and techniques are an important component of any field of study. The following chapter elucidates the various tools and techniques that are related to agricultural sciences; **Chapter 8** - This chapter focuses on those fields whose subject of study is similar to that of agricultural science. These subjects can guide agricultural scientists by offering ideas on moisture content, soil texture, drainage etc. Such ideas can greatly help to improve the growth of modified crops and plants. This chapter is a compilation of the various branches of agricultural science that form an integral part of the broader subject matter; **Chapter 9** - Pollination is defined as the natural process of plant reproduction. It is vital to the plant's survival and spread. The advance of technology has meant that it is possible to harness resources and elements that aid in the process of pollination. The chapter strategically encompasses and incorporates the major components and key concepts of pollination management, providing a complete understanding; **Chapter 10 -** Agricultural resources come to involve agricultural land as well as any tools, techniques and procedures that transform it into productive arable land. Discussed in this chapter are articles dealing with agricultural resources, agricultural machinery and fertilizer; **Chapter 11 -** Agricultural science has coherently recorded and applied the latest advancements that have taken place in the fields of agriculture, cultivation and irrigation. This chapter touches upon the major developments that have taken place in agricultural science in the last century-and-a-half.

At the end, I would like to thank all the people associated with this book devoting their precious time and providing their valuable contributions to this book. I would also like to express my gratitude to my fellow colleagues who encouraged me throughout the process.

Editor

Introduction to Agricultural Science

Agricultural science can be defined as the application of agricultural techniques in compliance with sustainable development practices. Agricultural science is deeply integrated with global food production and consumption. The practice of agricultural productivity and pest management is a key component of this discipline. This chapter is an overview of the subject matter incorporating all the major aspects of agricultural science.

Agricultural Science

Agricultural science is a broad multidisciplinary field of biology that encompasses the parts of exact, natural, economic and social sciences that are used in the practice and understanding of agriculture. (Veterinary science, but not animal science, is often excluded from the definition.)

Agriculture, Agricultural Science, and Agronomy

The three terms are often confused. However, they cover different concepts:

- Agriculture is the set of activities that transform the environment for the production of animals and plants for human use. Agriculture concerns techniques, including the application of agronomic research.

- Agronomy is research and development related to studying and improving plant-based crops.

Agricultural sciences include research and development on:

- Production techniques (e.g., irrigation management, recommended nitrogen inputs)

- Improving agricultural productivity in terms of quantity and quality (e.g., selection of drought-resistant crops and animals, development of new pesticides, yield-sensing technologies, simulation models of crop growth, in-vitro cell culture techniques)

- Minimizing the effects of pests (weeds, insects, pathogens, nematodes) on crop or animal production systems.

- Transformation of primary products into end-consumer products (e.g., production, preservation, and packaging of dairy products)

- Prevention and correction of adverse environmental effects (e.g., soil degradation, waste management, bioremediation)

- Theoretical production ecology, relating to crop production modeling

- Traditional agricultural systems, sometimes termed subsistence agriculture, which feed most of the poorest people in the world. These systems are of interest as they sometimes retain a level of integration with natural ecological systems greater than that of industrial agriculture, which may be more sustainable than some modern agricultural systems.

- Food production and demand on a global basis, with special attention paid to the major producers, such as China, India, Brazil, the USA and the EU.

- Various sciences relating to agricultural resources and the environment (e.g. soil science, agro-climatology); biology of agricultural crops and animals (e.g. crop science, animal science and their included sciences, e.g. ruminant nutrition, farm animal welfare); such fields as agricultural economics and rural sociology; various disciplines encompassed in agricultural engineering.

Agricultural Biotechnology

Agricultural biotechnology is a specific area of agricultural science involving the use of scientific tools and techniques, including genetic engineering, molecular markers, molecular diagnostics, vaccines, and tissue culture, to modify living organisms: plants, animals, and microorganisms.

Fertilizer

One of the most common yield reducers is because of fertilizer not being applied in slightly higher quantities during transition period, the time it takes the soil to rebuild its aggregates and organic matter. Yields will decrease temporarily because of nitrogen being immobilized in the crop residue, which can take a few months to several years to decompose, depending on the crop's C to N ratio and the local environment.

A local science

With the exception of theoretical agronomy, research in agronomy, more than in any other field, is strongly related to local areas. It can be considered a science of ecoregions, because it is closely linked to soil properties and climate, which are never exactly the same from one place to another. Many people think an agricultural production system relying on local weather, soil characteristics, and specific crops has to be studied locally. Others feel a need to know and understand production systems in as many areas as possible, and the human dimension of interaction with nature.

History of agricultural science

Agricultural science began with Gregor Mendel's genetic work, but in modern terms might be better dated from the chemical fertilizer outputs of plant physiological understanding in 18th-century Germany. In the United States, a scientific revolution in agriculture began with the Hatch Act of 1887, which used the term "agricultural science". The Hatch Act was driven by farmers' interest in knowing the constituents of early artificial fertilizer. The Smith-Hughes Act of 1917 shifted agricultural education back to its vocational roots, but the scientific foundation had been built. After 1906, public expenditures on agricultural research in the US exceeded private expenditures for the next 44 years.

Intensification of agriculture since the 1960s in developed and developing countries, often referred to as the Green Revolution, was closely tied to progress made in selecting and improving crops and animals for high productivity, as well as to developing additional inputs such as artificial fertilizers and phytosanitary products.

As the oldest and largest human intervention in nature, the environmental impact of agriculture in general and more recently intensive agriculture, industrial development, and population growth have raised many questions among agricultural scientists and have led to the development and emergence of new fields. These include technological fields that assume the solution to technological problems lies in better technology, such as integrated pest management, waste treatment technologies, landscape architecture, genomics, and agricultural philosophy fields that include references to food production as something essentially different from non-essential economic 'goods'. In fact, the interaction between these two approaches provide a fertile field for deeper understanding in agricultural science.

New technologies, such as biotechnology and computer science (for data processing and storage), and technological advances have made it possible to develop new research fields, including genetic engineering, agrophysics, improved statistical analysis, and precision farming. Balancing these, as above, are the natural and human sciences of agricultural science that seek to understand the human-nature interactions of traditional agriculture, including interaction of religion and agriculture, and the non-material components of agricultural production systems.

Prominent Agricultural Scientists

Norman Borlaug, father of the Green Revolution.

- Robert Bakewell

- Norman Borlaug

- Luther Burbank

- George Washington Carver

- René Dumont
- Sir Albert Howard
- Kailas Nath Kaul
- Justus von Liebig
- Jay Lush
- Gregor Mendel
- Louis Pasteur
- M. S. Swaminathan
- Jethro Tull
- Artturi Ilmari Virtanen
- Eli Whitney
- Sewall Wright

Agricultural Science and Agriculture Crisis

Agriculture sciences seek to feed the world's population while preventing biosafety problems that may affect human health and the environment. This requires promoting good management of natural resources and respect for the environment, and increasingly concern for the psychological wellbeing of all concerned in the food production and consumption system.

Economic, environmental, and social aspects of agriculture sciences are subjects of ongoing debate. Recent crises (such as avian influenza, mad cow disease and issues such as the use of genetically modified organisms) illustrate the complexity and importance of this debate.

Fields Or Related Disciplines

• Agricultural biotechnology	• Aquaculture
• Agricultural chemistry	• Biological engineering
• Agricultural diversification	o Genetic engineering
• Agricultural education	• Nematology
• Agricultural economics	• Microbiology
• Agricultural engineering	o Plant pathology
• Agricultural geography	o Range Management
• Agricultural philosophy	• Environmental science
• Agricultural marketing	• Entomology
• Agricultural soil science	• Food science
• Agrophysics	o Human nutrition

- Animal science
  - Animal breeding
  - Animal husbandry
  - Animal nutrition
- Agronomy
  - Botany
  - Theoretical production ecology
  - Horticulture
  - Plant breeding
  - Plant fertilization
- Irrigation and water management
- Soil science
  - Agrology
- Waste management
- Weed science

Principles and Practices of Agricultural Science

2

Agricultural practices vary according to location as well as the demand for food products in an area. This is why there are a variety of agricultural practices that exist and are utilised alongside each other. The topics discussed in the chapter are of great importance to broaden the existing knowledge on agricultural science.

Mechanised Agriculture

Mechanised agriculture is the process of using agricultural machinery to mechanise the work of agriculture, greatly increasing farm worker productivity. In modern times, powered machinery has replaced many jobs formerly carried out by manual labour or by working animals such as oxen, horses and mules.

A cotton picker at work. The first successful models were introduced in the mid-1940s and each could do the work of 50 hand pickers.

The history of agriculture contains many examples of tool use, such as the plough. Mechanization involves the use of an intermediate device between the power source and the work. This intermediate device usually transforms motion, such as rotary to linear, or provides some sort of mechanical advantage, such as speed increase or decrease or leverage.

Current mechanised agriculture includes the use of tractors, trucks, combine harvesters, airplanes (crop dusters), helicopters, and other vehicles. Modern farms even sometimes use computers in conjunction with satellite imagery and GPS guidance to increase yields.

Mechanisation was one of the large factors responsible for urbanization and industrial economies. Besides improving production efficiency, mechanisation encourages large scale production and improves the quality of farm produce. On the other hand, it displaces unskilled farm labor, causes environmental pollution, deforestation and erosion.

History

Jethro Tull's seed drill (ca. 1701) was a mechanical seed spacing and depth placing device that increased crop yields and saved seed. It was an important factor in the British Agricultural Revolution.

A reaper at Woolbrook, New South Wales

Since the beginning of agriculture threshing was done by hand with a flail, requiring a great deal of labor. The threshing machine, which was invented in 1794 but not widely used for several more decades, simplified the operation and allowed the use of animal power. Before the invention of the grain cradle (ca. 1790) an able bodied laborer could reap about one quarter acre of wheat in a day using a sickle. It was estimated that for each of Cyrus McCormick's horse pulled reapers (ca. 1830s) freed up five men for military service in the U.S. Civil War. Later innovations included raking and binding machines. By 1890 two men and two horses could cut, rake and bind 20 acres of wheat per day.

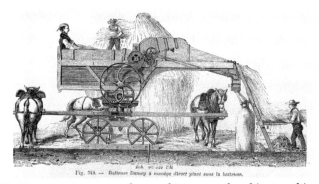

Threshing machine in 1881. Steam engines were also used to power threshing machines. Today both reaping and threshing are done with a combine harvester.

"Better and cheaper than horses" was the theme of many advertisements of the 1910s through 1930s.

In the 1880s the reaper and threshing machine were combined into the combine harvester. These machines required large teams of horses or mules to pull. Steam power was applied to threshing machines in the late 19th century. There were steam engines that moved around on wheels under their own power for supplying temporary power to stationary threshing machines. These were called *road engines,* and Henry Ford seeing one as a boy was inspired to build an automobile.

With internal combustion came the first modern tractors in the early 1900s, becoming more popular after the Fordson tractor (ca. 1917). At first reapers and combine harvesters were pulled by tractors, but in the 1930s self powered combines were developed. (*Link to a chapter on agricultural mechanisation in the 20th Century at reference*)

Advertising for motorized equipment in farm journals during this era did its best to compete against horse-drawn methods with economic arguments, extolling common themes such as that a tractor "eats only when it works", that one tractor could replace many horses, and that mechanisation could allow one man to get more work done per day than he ever had before. The horse population in the U.S. began to decline in the 1920s after the conversion of agriculture and transportation to internal combustion. Peak tractor sales in the U.S. were around 1950. In addition to saving labor, this freed up much land previously used for supporting draft animals. The greatest period of growth in agricultural productivity in the U.S. was from the 1940s to the 1970s, during which time agriculture was benefiting from internal combustion powered tractors and combine harvesters, chemical fertilizers and the green revolution.

Although farmers of corn, wheat, soy, and other commodity crops had replaced most of their workers with harvesting machines and combines enabling them to efficiently cut and gather grains, growers of produce continued to rely on human pickers to avoid the bruising of the product in order to maintain the blemish-free appearance demanded of consumers. The continuous supply of illegal workers from Latin America that were willing to harvest the crops for low wages further suppressed the need for mechanization. As the number of illegal workers has continued to decline since reaching its peak in 2007 due to increased border patrols and an improving Mexican economy, the industry is increasing the use of mechanization. Proponents argue that mechanization will boost productivity and help to maintain low food prices while farm worker advocates assert that it will eliminate jobs and will give an advantage to large growers who are able to afford the required equipment.

Current Status of Future Applications

Asparagus Harvesting

Asparagus are presently harvested by hand with labor costs at 71% of production costs and 44% of selling costs. Asparagus is a difficult crop to harvest since each spear matures at a different speed making it difficult to achieve a uniform harvest. A prototype asparagus harvesting machine - using a light-beam sensor to identify the taller spears - is expected to be available for commercial use.

Blueberry Harvesting

Mechanization of Maine's blueberry industry has reduced the number of migrant workers required from 5,000 in 2005 to 1,500 in 2015 even though production has increased from 50-60 million pounds per year in 2005 to 90 million pounds in 2015.

Chili Pepper Harvesting

As of 2014, prototype chili pepper harvesters are being tested by New Mexico State University. The New Mexico green chile crop is currently hand-picked entirely by field workers as chili pods tend to bruise easily. The first commercial application commenced in 2015. The equipment is expected to increase yield per acre and help to offset a sharp decline in acreage planted due to the lack of available labor and drought conditions.

Orange Harvesting

As of 2010, approximately 10% of the processing orange acreage in Florida is harvested mechanically. Mechanisation has progressed slowly due to the uncertainty of future economic benefits due to competition from Brazil and the transitory damage to orange trees when they are harvested.

Raisin Harvesting

As of 2007, mechanised harvesting of raisins is at 45%; however the rate has slowed due to high raisin demand and prices making the conversion away from hand labor less urgent. A new strain of grape developed by the USDA that drys on the vine and is easily harvested mechanically is expected to reduce the demand for labor.

Strawberry Harvesting

Strawberries are a high cost-high value crop with the economics supporting mechanisation. In 2005, picking and hauling costs were estimated at $594 per ton or 51% of the total grower cost. However, the delicate nature of fruit make it an unlikely candidate for mechanisation in the near future. A strawberry harvester developed by Shibuya Seiki and unveiled in Japan in 2013 is able to pick a strawberry every eight seconds. The robot identifies which strawberries are ready to pick by using three separate cameras and then once identified as ready, a mechanized arm snips the fruit free and gently places it in a basket. The robot moves on rails between the rows of strawberries which are generally contained within elevated greenhouses. The machine costs 5 million yen. A new strawberry harvester made by Agrobot that will harvest strawberries on raised, hydroponic beds using 60 robotic arms is expected to be released in 2016.

Tomato Harvesting

Mechanical harvesting of tomatoes started in 1965 and as of 2010, nearly all processing tomatoes are mechanically harvested. As of 2010, 95% of the U.S. processed tomato crop is produced in California. Although fresh market tomatoes have substantial hand harvesting costs (in 2007, the costs of hand picking and hauling were $86 per ton which is 19% of total grower cost), packing and selling costs were more of a concern (at 44% of total grower cost) making it likely that cost saving efforts would be applied there.

According to a 1977 report by the California Agrarian Action Project, during the summer of 1976 in California, many harvest machines had been equipped with a photo-electric scanner that sorted out green tomatoes among the ripe red ones using infrared lights and color sensors. It worked in lieu of 5,000 hand harvesters causing displacement of innumerable farm laborers as well as wage cuts and shorter work periods. Migrant workers were hit the hardest. To withstand the rigor of the machines, new crop varieties were bred to match the automated pickers. UC Davis Professor G.C. Hanna propagated a thick-skinned tomato called VF-145. But even still, millions were damaged with impact cracks and university breeders produced a more tougher and juiceless "square round" tomato. Small farms were of insufficient size to obtain financing to purchase the equipment and within 10 years, 85% of the state's 4,000 cannery tomato farmers were out of the business. This led to a concentrated tomato industry in California that "now packed 85% of the nation's tomato products". The monoculture fields fostered rapid pest growth, requiring the use of "more than four million pounds of pesticides each year" which greatly affected the health of the soil, the farm workers, and possibly the consumers.

Urban Agriculture

Urban agriculture, urban farming or urban gardening is the practice of cultivating, processing, and distributing food in or around a village, town, or city. Urban agriculture can also involve animal husbandry, aquaculture, agroforestry, urban beekeeping, and horticulture. These activities occur in peri-urban areas as well, and peri-urban agriculture may have different characteristics.

An urban farm in Chicago

Urban agriculture can reflect varying levels of economic and social development. In the global north, it often takes the form of a social movement for sustainable communities, where organic growers, 'foodies,' and 'locavores' form social networks founded on a shared ethos of nature and community holism. These networks can evolve when receiving formal institutional support, becoming integrated into local town planning as a 'transition town' movement for sustainable urban development. In the developing south, food security, nutrition, and income generation are key motivations for the practice. In either case, more direct access to fresh vegetables, fruits, and meat products through urban agriculture can improve food security and food safety.

History

Huerto (vegetable garden or orchard) Romita, organization dedicated to urban agriculture located in the La Romita section of Colonia Roma, Mexico City

Community wastes were used in ancient Egypt to feed urban farming. In Machu Picchu, water was conserved and reused as part of the stepped architecture of the city, and vegetable beds were designed to gather sun in order to prolong the growing season. Allotment gardens came up in

Germany in the early 19th century as a response to poverty and food insecurity. Victory gardens sprouted during WWI and WWII and were fruit, vegetable, and herb gardens in US, Canada, and UK. This effort was undertaken by citizens to reduce pressure on food production that was to support the war effort. Community gardening in most communities are open to the public and provide space for citizens to cultivate plants for food or recreation. A community gardening program that is well-established is Seattle's P-Patch. The grass roots permaculture movement has been hugely influential in the renaissance of urban agriculture throughout the world. The Severn Project in Bristol was started in 2010 for £2500 and provides 34 tons of produce per year, employing people from disadvantaged backgrounds

The idea of supplemental food production beyond rural farming operations and distant imports is not new and has been used during war times and the Great Depression when food shortage issues arose. As early as 1893, citizens of a depression-struck Detroit were asked to use any vacant lots to grow vegetables. They were nicknamed Pingree's Potato Patches after the mayor, Hazen S. Pingree, who came up with the idea. He intended for these gardens to produce income, food supply, and even boost independence during times of hardship.

During the first World War, president Woodrow Wilson called upon all American citizens to utilize any available open space for food growth, seeing this as a way to pull them out of a potentially damaging situation. Because most of Europe was consumed with war, they were unable to produce sufficient food supplies to be shipped to the U.S., and a new plan was implemented with the intent to feed the U.S. and even supply a surplus to other countries in need. By the year 1919, over 5 million plots were growing food and over 500 million pounds of produce was harvested. A very similar practice came into use during the Great Depression that provided a purpose, a job, and food to those who would otherwise be without anything during such harsh times. In this case, these efforts helped to raise spirits socially as well as to boost economic growth. Over 2.8 million dollars worth of food was produced from the subsistence gardens during the Depression. By the time of the Second World War, the War/Food Administration set up a National Victory Garden Program that set out to systematically establish functioning agriculture within cities. With this new plan in action, as many as 5.5 million Americans took part in the victory garden movement and over 9 million pounds of fruit and vegetables were grown a year, accounting for 44% of U.S.-grown produce throughout that time.

A tidy front yard flower and vegetable garden in Aretxabaleta, Spain

In 2010, New York City saw the building and opening of the world's largest privately owned and operated rooftop farm, followed by an even larger location in 2012. Both were a result of municipal programs such as The Green Roof Tax Abatement Program. and Green Infrastructure Grant Program

With its past success in mind and with modern technology, urban agriculture today can be something to help both developed and developing nations.

Perspectives

A vegetable garden in the square in front of the train station in Ezhou, China

Resource and Economic

The Urban Agriculture Network has defined urban agriculture as:

[A]n industry that produces, processes, and markets food, fuel, and other outputs, largely in response to the daily demand of consumers within a town, city, or metropolis, on many types of privately and publicly held land and water bodies found throughout intra-urban and peri-urban areas. Typically urban agriculture applies intensive production methods, frequently using and re-using natural resources and urban wastes, to yield a diverse array of land-, water-, and air-based fauna and flora, contributing to the food security, health, livelihood, and environment of the individual, household, and community.

Environmental

The Council on Agriculture, Science and Technology (CAST) defines urban agriculture to include aspects of environmental health, remediation, and recreation:

Urban agriculture is a complex system encompassing a spectrum of interests, from a traditional core of activities associated with the production, processing, marketing, distribution, and consumption, to a multiplicity of other benefits and services that are less widely acknowledged and documented. These include recreation and leisure; economic vitality and business entrepreneurship, individual health and well-being; community health and well being; landscape beautification; and environmental restoration and remediation.

Modern planning and design initiatives are often more responsive to this model of urban agriculture because it fits within the current scope of sustainable design. The definition allows for a multitude of interpretations across cultures and time. Frequently it is tied to policy decisions to build sustainable cities.

Food Security

Access to nutritious food, both economically and geographically, is another perspective in the effort to locate food and livestock production in cities. With the tremendous influx of world population to urban areas, the need for fresh and safe food is increased. The Community Food Security Coalition (CFSC) defines food security as:

All persons in a community having access to culturally acceptable, nutritionally adequate food through local, non-emergency sources at all times.

Areas faced with food security issues have limited choices, often relying on highly processed fast food or convenience store foods that are high in calories and low in nutrients, which may lead to elevated rates of diet-related illnesses such as diabetes. These problems have brought about the concept of food justice which Alkon and Norgaard (2009; 289) explain is, "places access to healthy, affordable, culturally appropriate food in the contexts of institutional racism, racial formation, and racialized geographies.... Food justice serves as a theoretical and political bridge between scholarship and activism on sustainable agriculture, food insecurity, and environmental justice."

Impact

A sprouting glass jar with mung beans in it.

Economic

Urban and peri-urban agriculture (UPA) expands the economic base of the city through production, processing, packaging, and marketing of consumable products. This results in an increase in entrepreneurial activities and the creation of jobs, as well as reducing food costs and improving quality. UPA provides employment, income, and access to food for urban populations, which helps to relieve chronic and emergency food insecurity. Chronic food insecurity refers to less affordable food and growing urban poverty, while emergency food insecurity relates to breakdowns in the chain of food distribution. UPA plays an important role in making food more affordable and in providing emergency supplies of food. Research into market values for produce grown in urban gardens has attributed to a community garden plot a median yield value of between approximately $200 and $500 (US, adjusted for inflation).

Social

The needs of urban landscaping can be combined with those of suburban livestock farmers. (Kstovo, Russia)

There are many social benefits that have emerged from urban agricultural practices, such as improved overall social and emotional well-being, improved health and nutrition, increased income, employment, food security within the household, and community social life. Urban agriculture can have a large impact on the social and emotional well-being of individuals. Individuals report to have decreased levels of stress and better overall mental health when they have opportunities to interact with nature through a garden. Urban gardens are thought to be relaxing and calming, and offer a space of retreat in densely populated urban areas

UA can have an overall positive impact on community health, which directly impacts individuals social and emotional well-being. There have been many documented cases in which community gardens lead to improved social relationships, increased community pride, and overall community improvement and mobilization (2). This improvement in overall community health can also be connected to decreased levels of crime and suicide rates

Urban gardens are often places that facilitate positive social interaction, which also contributes to overall social and emotional well-being. Many gardens facilitate the improvement of social networks within the communities that they are located. For many neighborhoods, gardens provide a "symbolic focus," which leads to increased neighborhood pride

When individuals come together around UA, physical activity levels are often increased. Everything that is involved in starting and maintaining a garden, from turning the soil to digging holes, contributes to an individual's physical activity. Many state that working in agriculture is much more interesting and fulfilling than going to the gym, and that it makes getting exercise "fun." In addition to the exercise that individuals receive while actually working in gardens, many people say that the majority of the exercise they receive through urban agriculture is actually getting to the gardens—many people either walk or ride their bike to the sites, which provides many physical benefits.

UPA can be seen as a means of improving the livelihood of people living in and around cities. Taking part in such practices is seen mostly as informal activity, but in many cities where inadequate, unreliable, and irregular access to food is a recurring problem, urban agriculture has been a positive response to tackling food concerns. Due to the food security that comes with UA, a feelings of independence and empowerment often arise. The ability to produce and grow food for oneself has also been reported to improve levels of self-esteem or of self-efficacy. Households and small communities take advantage of vacant land and contribute not only to their household food needs but also the needs of their resident city. The CFSC states that:

Community and residential gardening, as well as small-scale farming, save household food dollars. They promote nutrition and free cash for non-garden foods and other items. As an example you can raise your own chickens on an urban farm and have fresh eggs for only $0.44 per dozen.

This allows families to generate larger incomes selling to local grocers or to local outdoor markets, while supplying their household with proper nutrition of fresh and nutritional produce.

Some community urban farms can be quite efficient and help women find work, who in some cases are marginalized from finding employment in the formal economy. Studies have shown that participation from women have a higher production rate, therefore producing the adequate amount for household consumption while supplying more for market sale.

As most UA activities are conducted on vacant municipal land, there have been rising concerns about the allocation of land and property rights. The IDRC and the FAO have published the Guidelines for Municipal Policymaking on Urban Agriculture, and are working with municipal governments to create successful policy measures that can be incorporated in urban planning.

Energy Efficiency

The current industrial agriculture system is accountable for high energy costs for the transportation of foodstuffs. According to a study by Rich Pirog, the associate director of the Leopold Center for Sustainable Agriculture at Iowa State University, the average conventional produce item travels 1,500 miles (2,400 km), using, if shipped by tractor-trailer, 1 US gallon (3.8 l; 0.83 imp gal) of fossil fuel per 100 pounds (45 kg). The energy used to transport food is decreased when urban agriculture can provide cities with locally grown food. Pirog found that traditional, non-local, food distribution system used 4 to 17 times more fuel and emitted 5 to 17 times more CO_2 than the local and regional transport.

Edible Oyster Mushrooms growing on used coffee grounds

Similarly, in a study by Marc Xuereb and Region of Waterloo Public Health, they estimated that switching to locally grown food could save transport-related emissions equivalent to nearly 50,000 metric tons of CO_2, or the equivalent of taking 16,191 cars off the road.

A windowfarm, incorporating discarded plastic bottles into pots for hydroponic agriculture in urban windows

Carbon Footprint

As mentioned above, the energy-efficient nature of urban agriculture can reduce each city's carbon footprint by reducing the amount of transport that occurs to deliver goods to the consumer.

Also these areas can act as carbon sinks offsetting some of carbon accumulation that is innate to urban areas, where pavement and buildings outnumber plants. Plants absorb atmospheric carbon dioxide (CO_2) and release breathable oxygen (O_2) through photosynthesis. The process of Carbon Sequestration can be further improved by combining other agriculture techniques to increase removal from the atmosphere and prevent release of CO_2 during harvest time. However, this process relies heavily on the types of plants selected and the methodology of farming. Specifically, choosing plants that do not lose their leaves and remain green all year can increase the farms ability to sequester carbon.

Reduction in Ozone and Particulate Matter

The reduction in ozone and other particulate matter can benefit human health. Reducing these particulates and ozone gases could reduce mortality rates in urban areas along with increase the health of those living in cities. Just to give one example, in the article "Green roofs as a means of pollution abatement," the author argues that a rooftop containing 2000 m² of uncut grass has the potential to remove up to 4000 kg of particulate matter. According to the article, only one square meter of green roof is needed to offset the annual particulate matter emissions of a car.

Soil Decontamination

Vacant urban lots are often victim to illegal dumping of hazardous chemicals and other wastes. They are also liable to accumulate standing water and "grey water", which can be dangerous to public health, especially left stagnant for long periods. The implementation of urban agriculture in these vacant lots can be a cost-effective method for removing these chemicals. In the process known as Phytoremediation, plants and the associated microorganisms are selected for their chemical ability to degrade, absorb, convert to an inert form, and remove toxins from the soil. Several chemicals can be targeted for removal including heavy metals (e.g. Mercury and lead) inorganic compounds (e.g. Arsenic and Uranium), and organic compounds (e.g. petroleum and chlorinated compounds like PBC's).

Phytoremeditation is both an environmentally friendly, cost-effective, and energy-efficient measure to reduce pollution. Phytoremediation only costs about $5–$40 per ton of soil being decontaminated. Implementation of this process also reduces the amount of soil that must be disposed of in a hazardous waste landfill.

Urban agriculture as a method to mediate chemical pollution can be effective in preventing the spread of these chemicals into the surrounding environment. Other methods of remediation often disturb the soil and force the chemicals contained within it into the air or water. Plants can be used as a method to remove chemicals and also to hold the soil and prevent erosion of contaminated soil decreasing the spread of pollutants and the hazard presented by these lots.

Noise Pollution

Large amounts of noise pollution not only lead to lower property values and high frustration, they can be damaging to human hearing and health. In the study "Noise exposure and public health," they argue that exposure to continual noise is a public health problem. They cite examples of the detriment of continual noise on humans to include: "hearing impairment, hypertension and ischemic heart disease, annoyance, sleep disturbance, and decreased school performance." Since most roofs or vacant lots consist of hard flat surfaces that reflect sound waves instead of absorb them, adding plants that can absorb these waves has the potential to lead to a vast reduction in noise pollution.

Nutrition and Quality of Food

Daily intake of a variety of fruits and vegetables is linked to a decreased risk of chronic diseases including diabetes, heart disease and cancer. Urban agriculture is associated with increased consumption of fruits and vegetables which decreases risk for disease and can be a cost-effective way to provide citizens with quality, fresh produce in urban settings.

People are more likely to try new vegetables when they take an active role in the planting and cultivation of an urban garden. Produce from urban gardens can be perceived to be more flavorful and desirable than store bought produce which may also lead to a wider acceptance and higher intake. A Flint, Michigan study found that those participating in community gardens consumed fruits and vegetables 1.4 more times per day and were 3.5 times more likely to consume fruits or vegetables at least 5 times daily. Garden-based education can also yield nutritional benefits in children. An Idaho study reported a positive association between school gardens and increased intake of fruit, vegetables, vitamin A, vitamin C and fiber among sixth graders.

Urban gardening improves dietary knowledge. Inner city youth of Minneapolis/St. Paul, Minnesota who were part of a community garden intervention were better able to communicate specific nutritional benefits of fruits and vegetables on the body than those who had not participated in a community garden. Community gardeners were also found to consume fewer sweet foods and drinks in a Philadelphia study.

The nutrient content of produce from an urban garden may be higher due to decrease in time between production and consumption. A 30-50% nutrient loss can happen in the 5–10 days it takes to travel from farm-to-table. Harvesting fruits and vegetables initiates the enzymatic process of nutrient degradation which is especially detrimental to water soluble vitamins such as ascorbic acid and thiamin. The process of blanching produce in order to freeze or can reduces nutrient content slightly but not nearly as much as the amount of time spent in storage. Harvesting produce from one's own community garden cuts back on storage times significantly.

Urban agriculture also provides quality nutrition for low income households. Studies show that every $1 invested in a community garden yields $6 worth of vegetables, if labor is not considered a factor in investment. Many urban gardens reduce the strain on food banks and other emergency food providers by donating shares of their harvest and provide fresh produce in areas that otherwise might be food deserts. The supplemental nutrition program Women, Infants and Children (WIC) as well as the Supplemental Nutrition Assistance Program (SNAP) have partnered with several urban gardens nationwide to improve the accessibility to produce in exchange for a few hours of volunteer gardening work.

Economy of Scale

Using high-density urban farming, as for instance with vertical farms or stacked greenhouses, many environmental benefits can be achieved on a city-wide scale that would be impossible otherwise. These systems do not only provide food, but also produce potable water from waste water, and can recycle organic waste back to energy and nutrients. At the same time, they can reduce food-related transportation to a minimum while providing fresh food for large communities in almost any climate.

Health Inequalities and Food Justice

A 2009 report by the USDA, determined that "Evidence is both abundant and robust enough for us to conclude that Americans living in low-income and minority areas tend to have poor access to healthy food", and that the "structural inequalities" in these neighborhoods "contribute to inequalities in diet and diet-related outcomes". These diet related outcomes, including obesity and diabetes, have become epidemic in low-income urban environments in the United States. Although the definition and methods for determining "food deserts" have varied, studies indicate that, at least in the United States, there are racial disparities in the food environment. Thus using the definition of environment as the place where people live, work, play and pray, food disparities become an issue of environmental justice. This is especially true in American inner-cities where a history of racist practices have contributed to the development of food deserts in the low-income, minority areas of the urban core. The issue of inequality is so integral to the issues of food access and health that the Growing Food & Justice for All Initiative was founded with the mission of "dismantling racism" as an integral part of creating food security.

Environmental Justice

Urban agriculture may advance environmental justice and food justice for communities living in food deserts. First, urban agriculture may reduce racial and class disparities in access to healthy food. When urban agriculture leads to locally grown fresh produce sold at affordable prices in food deserts, access to healthy food is not just available for those who live in wealthy areas, thereby leading to greater equity in rich and poor neighborhoods.

Improved access to food through urban agriculture can also help alleviate psychosocial stresses in poor communities. Community members engaged in urban agriculture improve local knowledge about healthy ways to fulfill dietary needs. Urban agriculture can also better the mental health of community members. Buying and selling quality products between local producers and consumers allows community members to support one another, which may reduce stress. Thus, urban agriculture can help improve conditions in poor communities, where residents undergo higher levels of stress due to hopeless caused by a lack of control over the quality of their lives.

Urban agriculture may improve the livability and built environment in communities that lack supermarkets and other infrastructure due to the presence of high unemployment caused by deindustrialization. Urban farmers who follow sustainable agriculture methods can not only help to build local food system infrastructure, but can also contribute to improving local air, and water and soil quality. When agricultural products are produced locally within the community, they do not need to be transported, which reduces CO_2 emission rates and other pollutants that contribute to high rates of asthma in lower socioeconomic areas. Sustainable urban agriculture can also promote worker protection and consumer rights. For example, communities in New York, Illinois, and Richmond, Virginia have demonstrated improvements to their local environments through urban agricultural practices.

However, urban agriculture can also present urban growers with health risks if the soil used for urban farming is contaminated. Although local produce is often believed to be clean and healthy, many urban farmers ranging from New York urban farmer Frank Meushke to Presidential First Lady Michelle Obama have found their produce contained high levels of lead, due to soil contamination, which is harmful to human health when consumed. The soil contaminated with high lead levels often originates from old house paint containing lead, vehicle exhaust, or atmospheric deposition. Without proper education on the risks of urban farming and safe practices, urban consumers of urban agricultural produce may face additional health related issues

Implementation

A small urban farm in Amsterdam

Rooftop urban farming at the Food Roof Farm in downtown St. Louis, MO

Creating a community-based infrastructure for urban agriculture means establishing local systems to grow and process food and transfer it from farmer (producer) to consumer.

To facilitate food production, cities have established community-based farming projects. Some projects have collectively tended community farms on common land, much like that of eighteenth-century Boston Common. One such community farm is the Collingwood Children's Farm in Melbourne, Australia. Other community garden projects use the allotment garden model, in which gardeners care for individual plots in a larger gardening area, often sharing a tool shed and other amenities. Seattle's P-Patch gardens use this model, as did the South Central Farm in Los Angeles and the Food Roof Farm in St. Louis. Independent urban gardeners also grow food in individual yards and on roofs. Garden sharing projects seek to pair producers with land, typically, residential yard space. Roof gardens allow for urban dwellers to maintain green spaces in the city without having to set aside a tract of undeveloped land. Rooftop farms allow otherwise unused industrial roofspace to be used productively, creating work and profit. Projects around the world seek to enable cities to become 'continuous productive landscapes' by cultivating vacant urban land and temporary or permanent kitchen gardens.

Tomato plants growing in a pot farming alongside a small house in New Jersey in fifteen garbage cans filled with soil, grew over 700 tomatoes during the summer of 2013.

Food processing on a community level has been accommodated by centralizing resources in community tool sheds and processing facilities for farmers to share. The Garden Resource Program Collaborative based in Detroit has cluster tool banks. Different areas of the city have toolbanks where resources like tools, compost, mulch, tomato stakes, seeds, and education can be shared and distributed with the gardeners in that cluster. Detroit's Garden Resource Program Collaborative also strengthens their gardening community by providing to their members transplants; education on gardening, policy, and food issues; and by building connectivity between gardeners through workgroups, potlucks, tours, field trips, and cluster workdays.

Farmers' markets, such as the farmers' market in Los Angeles, provide a common land where farmers can sell their product to consumers. Large cities tend to open their farmers markets on the weekends and one day in the middle of the week. For example, the farmers' market of Boulevard Richard-Lenoir in Paris, France, is open on Sundays and Thursdays. However, to create a consumer dependency on urban agriculture and to introduce local food production as a sustainable career for farmers, markets would have to be open regularly. For example, the Los Angeles Farmers' Market is open seven days a week and has linked several local grocers together to provide different food products. The market's central location in downtown Los Angeles provides the perfect interaction for a diverse group of sellers to access their consumers.

Queensland, Australia

In Queenslands many people have started a trend of urban farming both utilizing Aquaponics and self watering containers. One man by the name of Rob Bob created Rob's Bits out the Back, Urban farming channel on YouTube documenting his stories and helping others to utilize their urban setting for farming. He has his whole yard front and back made into a gardening paradise.

Cairo, Egypt

In Egypt, development of rooftop gardens began in the 1990s. In the early 1990s at Ain Shams University, a group of agriculture professors developed an initiative focused on growing organic vegetables to suit densely populated cities of Egypt. The initiative was applied on a small scale; until it was officially adopted in 2001, by the Food and Agriculture Organization (FAO).

Havana, Cuba

Due to the shortage of fuel during the crisis and therefore severe deficiencies in the transportation sector, a growing percentage of the agricultural production takes place in the so-called urban agriculture. In 2002, 35,000 acres (14,000 ha) of urban gardens produced 3,400,000 short tons (3,100,000 t) of food. In Havana, 90% of the city's fresh produce come from local urban farms and gardens. In 2003, more than 200,000 Cubans worked in the expanding urban agriculture sector.

Mumbai, India

Economic development in Mumbai brought a growth in population caused mainly by the migration of laborers from other regions of the country. The number of residents in the city increased more than twelve times in the last century. Greater Mumbai, formed by City Island and Salsette Island, is the largest city in India with a population of 16.4 million, according to data collected by the census of 2001. Mumbai is one of the densest cities in the world, 48,215 persons per km^2 and 16,082 per km^2 in suburban areas. In this scenario, urban agriculture seems unlikely to be put into practice since it must compete with real estate developers for the access and use of vacant lots. Alternative farming methods have emerged as a response to scarcity of land, water, and economic resources employed in UPA.

Dr. Doshi's city garden methods are revolutionary for being appropriate to apply in reduced spaces as terraces and balconies, even on civil construction walls, and for not requiring big investments in capital or long hours of work. His farming practice is purely organic and is mainly directed to domestic consumption. His gardening tools are composed of materials available in the local environment: sugarcane waste, polyethylene bags, tires, containers and cylinders, and soil. The containers and bags (open at both ends) are filled with the sugarcane stalks, compost, and garden soil, which make possible the use of minimal quantity of water if compared to open fields. Dr. Doshi states that solar energy can replace soil in cities. He also recommends the idea of chainplanting, or growing plants in intervals and in small quantities rather than at once and in large amounts. He has grown different types of fruit such as mangos, figs, guavas, bananas, and sugarcane stalks in his terrace of 1,200 sq ft (110 m²) in Bandra. The concept of city farming developed by Dr. Doshi consumes the entire household's organic waste. He subsequently makes the household self-sufficient in the provision of food: 5 kilograms (11 lb) of fruits and vegetables are produced daily for 300 days a year.

The main objectives of a pilot project at city farm at Rosary High School, Dockyard Road, were to promote economic support for street children, beautify the city landscape, supply locally produced organic food to urban dwellers (mainly those residing in slums), and to manage organic waste in a sustainable city. The project was conducted in the Rosary School, in Mumbai, with the participation of street children during 2004. A city farm was created in a terrace area of 400 sq ft (37 m²).

The participants were trained in urban farming techniques. The farm produced vegetables, fruits, and flowers. The idea has spread the concept of city farm to other schools in the city.

The Mumbai Port Trust (MBPT) central kitchen distributes food to approximately 3,000 employees daily, generating important amounts of organic disposal. A terrace garden created by the staff recycles ninety percent of this waste in the production of vegetables and fruits. Preeti Patil, who is the catering officer at the MBPT explains the purpose of the enterprise:

Mumbai Port Trust has developed an organic farm on the terrace of its central kitchen, which is an area of approximately 3,000 sq ft (280 m²). The activity of city farming was started initially to dispose of kitchen organic waste in an ecofriendly way. Staff members, after their daily work in the kitchen, tend the garden, which has about 150 plants.

Bangkok, Thailand

In early 2000, urban gardens were started under the direction of the NGO, Thailand Environment Institute (TEI), to help achieve the Bangkok Metropolitan Administrations (BMA) priority to 'green' Thailand. With a population of 12 million and 39% of the land in the city vacant due to rapid expansion of the 1960s–80s Bangkok is a test bed for urban gardens centered on community involvement. The two urban gardens initiated by TEI are in Bangkok Noi and Bangkapi and the main tasks were stated as:

- Teach members of the communities the benefits of urban green space.
- Create the social framework to plan, implement, and maintain the urban green space.
- Create a process of method to balance the needs of the community with the needs of the larger environmental concerns.

While the goals of the NGO are important in a global context, the community goals are being met through the work of forming the urban gardens themselves. In this sense, the creation, implementation, and maintenance of urban gardens is highly determined by the desires of the communities involved. However, the criteria by which TEI measured their success illustrates the scope of benefits to a community which practices urban agriculture. TEI's success indicators were:

- Establishing an Urban Green Plan
- Community Capacity Building
- Poverty Reduction
- Links with Government
- Developing a Model for Other Communities

Evan D.G. Fraser wrote in the article Urban Ecology in Bangkok Thailand that although the project was initiated to serve the environmental needs of the city it quickly illustrated the positive side effects of urban agriculture:

In many ways, the urban environment became a lens through which communities re-evaluated their own relationship with the city, the impact of urbanization in a global context, and how small groups can exert some control over the shape of their neighbourhoods.

China

Beijing's increase in land area from 4,822 square kilometres (1,862 sq mi) in 1956 to 16,808 square kilometres (6,490 sq mi) in 1958 led to the increased adoption of peri-urban agriculture. Such "suburban agriculture" led to more than 70% of non-staple food in Beijing, mainly consisting of vegetables and milk, to be produced by the city itself in the 1960s and 1970s. Recently, with relative food security in China, periurban agriculture has led to improvements in the quality of the food available, as opposed to quantity. One of the more recent experiments in urban agriculture is the Modern Agricultural Science Demonstration Park in Xiaotangshan.

Traditionally, Chinese cities have been known to mix agricultural activities within the urban setting. Shenzhen, once a small farming community, is now a fast-growing metropolis due to the Chinese government designation as an open economic zone. Due to large and growing population in China, the government supports urban self-sufficiency in food production. Shenzhen's village structure, sustainable methods, and new agricultural advancements initiated by the government have been strategically configured to supply food for this growing city.

The city farms are located about 10 kilometres (6.2 mi) from city center in a two-tier system. The first tier approached from city center produces perishable items. Located just outside these farms, hardier vegetables are grown such as potatoes, carrots, and onions. This system allows produce to be sold in city markets just a few short hours after picking.

Another impressive method used within Chinese agriculture and aquaculture practice is the mulberry-dyke fish-pond system, which is a response to waste recycling and soil fertility. This system can be described as:

Mulberry trees are grown to feed silkworms and the silkworm waste is fed to the fish in ponds. The fish also feed on waste from other animals, such as pigs, poultry, and buffalo. The animals in turn are given crops that have been fertilized by mud from the ponds. This is a sophisticated system as a continuous cycle of water, waste and food...with man built into the picture.

As population grows and industry advances, the city tries to incorporate potential agricultural growth by experimenting in new agricultural methods. The Fong Lau Chee Experimental Farm in Dongguan, Guangdong has worked with new agricultural advancements in lychee production. This farm was established with aspirations of producing large quantities and high quality lychees, by constantly monitoring sugar content, and their seeds. This research, conducted by local agricultural universities allows for new methods to be used with hopes of reaching the needs of city consumers.

However, due to increased levels of economic growth and pollution, some urban farms have become threatened. The government has been trying to step in and create new technological advancements within the agricultural field to sustain levels of urban agriculture.

"The city plans to invest 8.82 billion yuan in 39 agricultural projects, including a safe agricultural base, an agricultural high-tech park, agricultural processing and distribution, forestry, eco-agricultural tourism, which will form an urban agriculture with typical Shenzhen characteristics" in conjunction with this program the city is expected to expand the Buji Farm Produce Wholesale Market.

According to the Municipal Bureau of Agriculture, Forestry and Fishery the city will invest 600 million yuan on farms located around the city, with hopes of the farms to provide "60 percent of the meat, vegetables and aquatic products in the Shenzhen market".

There has also been an emerging trend of going green and organic as a response to pollution and pesticides used in farming practices. Vegetable suppliers are required to pass certain inspections held by the city's Agriculture Bureau before they can be sold as "green".

New York

In New York City, many low-income residents suffer from high rates of obesity and diabetes, and limited sources of fresh produce. The City and local nonprofit groups have been providing land, training and financial encouragement, but the impetus in urban farming has really come from the farmers, who often volunteer when their regular work day is done. In addition, the New York City Department of Environmental Protection offers a grant program for private property owners in combined sewer areas of New York City. The minimum requirement is to manage 1" of storm water runoff from the contributing impervious area. Eligible projects include green roofs, rooftop farms, and rainwater harvesting on private property in combined sewer areas. Because of this grant program, New York City now has the world's largest rooftop farms.

Some urban gardeners have used empty lots to start community or urban gardens. However, the soil must be tested for heavy contamination in city soil because of vehicle exhaust and remnants of old construction. The City also has a composting program, which is available to gardeners and farmers. One group, GreenThumb, provides free seedlings. Another program, the City Farms project operated by the nonprofit Just Food, offers courses on growing and selling food.

Two alternate means of growing are rooftop gardens and hydroponic (soil-less) growing. *The New York Times* wrote an article about one of Manhattan's first gardens which incorporate both these techniques. Another option urban gardeners have used is Farm-in-A-Box LLC, a company that provides hand-made, ready-to-use garden boxes to residents and schools.

California

In response to the recession of 2008, a coalition of community-based organizations, farmers, and academic institutions in California's Pomona Valley formed the Pomona Valley Urban Agriculture Initiative.

After the passage of the North American Free Trade Agreement, cheap grain from the United States flooded Mexico, driving peasant farmers off of their land. Many immigrated into the Pomona Valley and found work in the construction industry. With the 2008 recession, the construction industry also suffered in the region. It is unlikely to regain its former strength because of severe water shortages in this desert region as well as ongoing weakness in the local economy. These immigrants were dry land organic farmers in their home country by default since they did not have access to pesticides and petroleum-based fertilizers. Now, they found themselves on the border of two counties: Los Angeles County with a population of 10 million and almost no farmland, and San Bernardino County which has the worst access to healthy food in the state. In both counties, there is a growing demand for locally grown organic produce. In response to these conditions, Uncommon Good, a community-based nonprofit organization that works with immigrant farmer families, convened a forum which became the Urban Farmers Association. The Urban Farmers Association is the first organization of its kind for poor immigrant farmers in the Pomona Valley. Its goal is to develop opportunities for its members to support themselves and their families through urban agriculture. With Uncommon Good, it is a founding member of the Pomona Valley Urban Agriculture Initiative (PVUAI). The PVUAI is working with local colleges and universities to expand upon a food assessment survey that was done in the City of Pomona.

Oakland

Urban agriculture in West Oakland has taken a radical form that can be traced back to community gardening initiatives starting in the 1970s in the cities of Berkeley and Oakland, and the city's African-American heritage. Oakland's manufacturing industry attracted new residents during WWII. To reduce racial tension, the Oakland Housing Authority established housing projects for blacks in West Oakland and whites in East Oakland. With exclusionary covenants and redlining by banks, development capital was kept out of West Oakland while the African-American population had limited opportunities to rent or buy housing outside West Oakland.

The Black Panther Party (BPP) played a role in seeding urban agricultural practices in West Oakland. One of its social programs aimed to improve the access to healthy food for the city's black population by providing breakfast in local schools, churches, and community centers. A small amount of this food came from small local gardens planted by BPP members. According to Prof. Nathan McClintock, "The Panthers used gardening as a coping mechanism and a means of supplementing their diets, as a well as a means to strengthen community members engaged in the struggle against oppression." The People of Color Greening Network (PCGN) was created in the 1990s. The group planted in empty and vacant lots in West Oakland. In addition, schools around Alameda County began teaching basic gardening skills and food education.

In 1998, the city of Oakland's Mayor's office of Sustainability proposed a Sustainable Community Development Initiative towards sustainable development. Due to West Oakland's lack of access to nutritious and healthy food, other organizations including the PCGN and City Slicker Farms demanded the plan include strategies for creating a sustainable impact within the local food system. City Slicker Farms was originally founded in 2001 in response to the lack of access to fresh produce in West Oakland. Through land donations from local residents, a network of urban farms was created through the Community Market Farms Program, and in 2005 the organization established the Backyard Garden Program to aid West Oakland residents in growing their own food at home. This program now grows upwards of 30,000 lbs. of food each year.

In 2005, Mayor Jerry Brown signed the UN World Environment Day Urban Environmental Accords, pledging Oakland to become a more sustainable city by the year 2012. This gave rise to Oakland City Council Resolutions, such as No. 76980 and No. 80332 which helped develop a Food Policy Council. It has teamed up with the Health for Oakland's People & Environment (HOPE) Collaborative, which works to improve the health and wellness of Oakland's residents. In 2009 the Oakland Food Policy Council started to plan urban agriculture in Oakland.

Canada

Lufa Farms greenhouses are constructed on the rooftops of Greater Montreal.

Canada has a number of companies working on urban farm technology including Lufa Farms and Alterrus Systems. In Montreal, there are 97 community gardens which allow citizens space in each plot to grow their plants. The program has been in place since 1975, and is managed by the boroughs. Of the eighteen boroughs, some have a gardening instructor who visits the gardens regularly to give gardeners tips. The soil, a water supply, a space for tools, sand, fencing, and paint are all provided by the city.

UK

Todmorden is a village of 17,000 inhabitants in Yorkshire, United Kingdom with a successful urban agriculture model. The project, which began in 2008, has meant that food crops have been planted at forty locations throughout the village. The produce is all free, the work is done by volunteers, and passers-by and visitors are invited to pick and use the produce., Some Todmorden plots have been permission plots while others have been examples of guerilla gardening. All are "propaganda gardens" promoting locals to consider growing local, to eat seasonal, to consider the provenance of their food, and to enjoy fresh. There are food plots in the street, in the health centre car park, at the rail station, in the police station, in the cemetery, and in all the village schools.

Argentina

The city of Rosario (population: 1.3 million) has incorporated agriculture fully into its land use planning and urban development strategy. Its Land Use Plan 2007-2017 makes specific provision for the agricultural use of public land. Under its Metropolitan Strategic Plan 2008-2018, Rosario is building a "green circuit", passing through and around the city, consisting of family and community gardens, large-scale, commercial vegetable gardens and orchards, multifunctional garden parks, and "productive barrios", where agriculture is integrated into programmes for the construction of public housing and the upgrading of slums. In 2014, the green circuit consisted of more than 30 ha of land used to grow vegetables, fruit and medicinal and aromatic plants. The city has five garden parks – large, landscaped green areas covering a total of 72 ha of land, which are used for agriculture and for cultural, sports and educational activities.

Benefits

The benefits that UPA brings along to cities that implement this practice are numerous. The transformation of cities from only consumers of food to generators of agricultural products contributes to sustainability, improved health, and poverty alleviation.

- UPA assists to close the open loop system in urban areas characterized by the importation of food from rural zones and the exportation of waste to regions outside the city or town.

- Wastewater and organic solid waste can be transformed into resources for growing agriculture products: the former can be used for irrigation, the latter as fertilizer.

- Vacant urban areas can be used for agriculture production.

- Other natural resources can be conserved. The use of wastewater for irrigation improves water management and increases the availability of freshwater for drinking and household consumption.

- UPA can help to preserve bioregional ecologies from being transformed into cropland.

- Urban agriculture saves energy (e.g. energy consumed in transporting food from rural to urban areas).

- Local production of food also allows savings in transportation costs, storage, and in product loss, what results in food cost reduction.

- UPA improves the quality of the urban environment through greening and thus, a reduction in pollution.

- Urban agriculture also makes of the city a healthier place to live by improving the quality of the environment.

- UPA is a very efficient tool to fight against hunger and malnutrition since it facilitates the access to food by an impoverished sector of the urban population.

Poverty alleviation: It is known that a large part of the people involved in urban agriculture is the urban poor. In developing countries, the majority of urban agricultural production is for self-consumption, with surpluses being sold in the market. According to the FAO (Food and Agriculture Organization of the United Nations), urban poor consumers spend between 60 and 80 percent of their income on food, making them very vulnerable to higher food prices.

- UPA provides food and creates savings in household expenditure on consumables, thus increasing the amount of income allocated to other uses.

- UPA surpluses can be sold in local markets, generating more income for the urban poor.

Community centers and gardens educate the community to see agriculture as an integral part of urban life. The Florida House Institute for Sustainable Development in Sarasota, Florida, serves as a public community and education center in which innovators with sustainable, energy-saving ideas can implement and test them. Community centers like Florida House provide urban areas with a central location to learn about urban agriculture and to begin to integrate agriculture with the urban lifestyle.

Urban farms also are a proven effective educational tool to teach kids about healthy eating and meaningful physical activity.

Trade-offs

- Space is at a premium in cities and is accordingly expensive and difficult to secure.

- The utilization of untreated waste water for urban agricultural irrigation can facilitate the spread of waterborne diseases among the human population.

- Although studies have demonstrated improved air quality in urban areas related to the proliferation of urban gardens, it has also been shown that increasing urban pollution (related specifically to a sharp rise in the number of automobiles on the road), has led to an increase in insect pests, which consume plants produced by urban agriculture. It is believed that changes to the physical structure of the plants themselves, which have been correlated to increased levels of air pollution, increase plants' palatability to insect pests. Reduced yields within urban gardens decreases the amount of food available for human consumption.

- Studies indicate that the nutritional quality of wheat suffers when urban wheat plants are exposed to high nitrogen dioxide and sulfur dioxide concentrations. This problem is particularly acute in the developing world, where outdoor concentrations of sulfur dioxide are high and large percentages of the population rely upon urban agriculture as a primary source of food. These studies have implications for the nutritional quality of other staple crops that are grown in urban settings.

- Agricultural activities on land that is contaminated (with such metals as lead) pose potential risks to human health. These risks are associated both with working directly on contaminated land and with consuming food that was grown in contaminated soil.

Municipal greening policy goals can pose conflicts. For example, policies promoting urban tree canopy are not sympathetic to vegetable gardening because of the deep shade cast by trees. However, some municipalities like Portland, Oregon, and Davenport, Iowa are encouraging the implementation of fruit bearing trees (as street trees or as park orchards) to meet both greening and food production goals.

References

- Hounshell, David A. (1984), From the American System to Mass Production, 1800-1932: The Development of Manufacturing Technology in the United States, Baltimore, Maryland: Johns Hopkins University Press, ISBN 978-0-8018-2975-8, LCCN 83016269

- Wells, David A. (1891). Recent Economic Changes and Their Effect on Production and Distribution of Wealth and Well-Being of Society. New York: D. Appleton and Co. ISBN 0-543-72474-3.

- Constable, George; Somerville, Bob (2003). A Century of Innovation: Twenty Engineering Achievements That Transformed Our Lives, Chapter 7, Agricultural Mechanization. Washington, DC: Joseph Henry Press. ISBN 0-309-08908-5.

- Morales, Alfonso (2011). "Growing Food and Justice: Dismantling Racism through Sustainable Food Systems". In Alison Hope Alkon; Julian Agyeman. Cultivating Food Justice: Race, Class, and Sustainability. MIT Press. pp. 149–177. ISBN 9780262300223.

- Nathan McClintock (2011). "From Industrial Garden to Food Desert: Demarcated Devaluation in the Flatlands of Oakland, California". In Alison Hope Alkon; Julian Agyeman. *Cultivating Food Justice: Race, Class, and Sustainability. MIT Press. pp. 89–121. ISBN 9780262300223.*

- *Nordahl, Darrin (2009). Public Produce: The New Urban Agriculture. Washington DC: Island Press. ISBN 978-1-59726-588-1.*

- Wall Street Journal: "Robots Step Into New Planting, Harvesting Roles - Labor shortage spurs farmers to use robots for handling delicate tasks in the fresh-produce industry" By ILAN BRAT April 23, 2015

- *"No Hands Touch the Land: Automating California Farms" (PDF). California Agrarian Action Project: 20–28. July 1977. Retrieved 2015-04-25.*

- *"Pirog, R. and A. Benjamin. ""Checking the food odometer: Comparing food miles for local versus conventional produce sales to Iowa institutions"", Leopold Center for Sustainable Agriculture, 2003" (PDF). Leopold. iastate.edu. Retrieved 1 April 2013.*

- *USDA; Economic Research Service (June 2009). "Access to Affordable and Nutritious Food: Measuring and Understanding Food Deserts and Their Consequences: A Report to Congress.". Administrative Publication No. (AP-036): 160. Retrieved 28 March 2013.*

- *Raja, Samina; Changxing Ma; Pavan Yadav (2008). "Beyond Food Deserts: Measuring and Mapping Racial Disparities in Neighborhood Food Environments". Journal of Planning Education and Research. 27 (4): 469–482. doi:10.1177/0739456X08317461. Retrieved 28 March 2013.*

- *McClintock, Nathan (2012). "Assessing lead contamination at multiple scales in Oakland, California: Implications for urban agriculture and environmental justice". Applied Geography. 35: 460–473. doi:10.1016/j. apgeog.2012.10.001.*

3

Agricultural Cycle: An Overview

Agricultural practices related to the growth and harvesting of crops is discussed in this chapter. Careful attention is given to techniques related to weed control and plant breeding. The chapter on agricultural cycle offers an insightful focus, keeping in mind the complex subject matter.

Agricultural Cycle

The agricultural cycle is the annual cycle of activities related to the growth and harvest of a crop. These activities include loosening the soil, seeding, special watering, moving plants when they grow bigger, and harvesting, among others.

The main steps for agricultural practices include preparation of soil, sowing, adding, manure and fertilizers, irrigation, harvesting and storage.

Seeding

The fundamental factor in the process of seeding is dependent on the properties of both seed and the soil it is being planted in. The prior step associated with seeding is crop selection, which mainly consists of two techniques: sexual and asexual. Asexual technique include all forms of vegetative process such as budding, grafting and layering. Sexual technique involves growing of the plant from a seed. Grafting is referred to as the artificial method of propagation in which parts of plants are joined together, in order to make them bind together and continue growing as one plant. Grafting is mainly applied to two parts of the plant: the dicot and the gymnosperms due to the presence of vascular cambium between the plant tissues: xylem and phloem. A grafted plant consists of two parts: first rootstock, which is the lower part of the plants that comprises roots and the lowest part of the shoot. Second the branches and primary stem,which consists of the upper and main part of the shoot which gradually develops into a fully nourished plant. Budding, is another form of asexual reproduction in which new plant develops from a productive objective source of the parent plant. It is a method in which a bud of the plant is joined onto the stem of another plant. The plant in which the bud is being implanted in, eventually develops into a replica of the parent plant. The new plant can either divert its ways into forming an independent plant, however in numerous cases the may remain attached and form various accumulations.

Germination

Germination is a process by which the seed develops into a seedling. The vital conditions necessary for this process are water, air, temperature, energy, viability and enzymes. If any of these conditions are absent, the process cannot undergo successfully. Germination is also known as sprouting; it is also considered as the first sign of life shown by a seed.

Pollination

The process of pollination refers to the transfer of pollen to the female organs of the plant. Optimum factor for ideal pollination are: relative humidity rate of 50-70% and temperature of 24.4 degrees Celsius. If the humidity rate is higher than 90%, the pollen would not shed. Increasing air circulation is a favourable method of keeping humidity levels under control.

Irrigation

Irrigation is the process of artificially applying water to soil to allow plant growth. This term is preferably used when large amounts of water is applied to dry, arid regions in order to facilitate plant growth. The process of irrigation not only increases the growth rate of the plant bust also increments the yield amount. In temperate and tropical areas rainfall and snowfall are the main suppliers of irrigation water, but in dry places with unfavourable weather conditions, groundwater serves as an essential source. Groundwater collects in basins made up of gravel and aquifers which are water-holding rocks.Dams also act as an essential distributive source of irrigation water. Underground wells also play an important role in storing water for irrigation, specifically in America and Arizona. Water and debris from streams filled by water accumulated during storms, also collects into underground basins. There are two types of irrigation techniques: spray irrigation and drip irrigation. Drip irrigation is regarded more efficient as less water evaporated as in spray irrigation.

Harvest

Harvesting agriculture in Volgograd Oblast, Russia

Sugar beet harvester. Baden-Wurttemberg, Germany.

Harvesting is the process of gathering a ripe crop from the fields. *Reaping* is the cutting of grain or pulse for harvest, typically using a scythe, sickle, or reaper. On smaller farms with minimal mechanization, harvesting is the most labor-intensive activity of the growing season. On large mechanized farms, harvesting utilizes the most expensive and sophisticated farm machinery, such as the combine harvester. The term "harvesting" in general usage may include immediate postharvest handling, including cleaning, sorting, packing, and cooling.

Rye harvest on Gotland, Sweden, 1900–1910.

The completion of harvesting marks the end of the growing season, or the growing cycle for a particular crop, and the social importance of this event makes it the focus of seasonal celebrations such as harvest festivals, found in many religions.

Etymology

"Harvest", a noun, came from the Old English word *hærfest*, meaning "autumn" (the season), "harvest-time", or "August". (It continues to mean "autumn" in British dialect, and "season of gathering crops" generally.) "The harvest" came to also mean the activity of reaping, gathering, and storing grain and other grown products during the autumn, and also the grain and other grown products themselves. "Harvest" was also verbified: "To harvest" means to reap, gather, and store the harvest (or the crop). People who harvest and equipment that harvests are harvesters; while they do it, they are harvesting.

Crop Failure

Crop failure (also known as harvest failure) is an absent or greatly diminished crop yield relative to expectation, caused by the plants being damaged, killed, or destroyed, or affected in some way that they fail to form edible fruit, seeds, or leaves in their expected abundance.

Crop failures can be caused by catastrophic events such as plant disease outbreaks, heavy rainfall, volcanic eruptions, storms, floods, or drought, or by slow, cumulative effects of soil degradation, too-high soil salinity, erosion, desertification, usually as results of drainage, overdrafting (for irrigation), overfertilization, or overexploitation.

In history, crop failures and subsequent famines have triggered human migration, rural exodus, etc.

The proliferation of industrial monocultures, with their reduction in crop diversity and dependence on heavy use of artificial fertilizers and pesticides, has led to overexploited soils that are nearly incapable of regeneration. Over years, unsustainable farming of land degrades soil fertility and diminishes crop yield. With a steadily growing world population and local overpopulation, even slightly diminishing yields are already the equivalent to a partial harvest failure.

Other uses

Harvesting commonly refers to grain and produce, but also has other uses. Fishing and logging are also referred to as harvesting. The term harvest is also used in reference to harvesting grapes for wine. Within the context of irrigation, *water harvesting* refers to the collection and run-off of rainwater for agricultural or domestic uses. Instead of *harvest*, the term *exploit* is also used, as in exploiting fisheries or water resources. *Energy harvesting* is the process of capturing and storing energy (such as solar power, thermal energy, wind energy, salinity gradients, and kinetic energy) that would otherwise go unexploited. *Body harvesting*, or *cadaver harvesting*, is the process of collecting and preparing cadavers for anatomical study. In a similar sense, *organ harvesting* is the removal of tissues or organs from a donor for purposes of transplanting.

Harvesting or *Domestic Harvesting* in Canada refers to hunting, fishing, and plant gathering by First Nations, Métis, and Inuit in discussions of aboriginal or treaty rights. For example, in the Gwich'in Comprehensive Land Claim Agreement, "Harvesting means gathering, hunting, trapping or fishing..." Similarly, in the Tlicho Land Claim and Self Government Agreement, "'Harvesting' means, in relation to wildlife, hunting, trapping or fishing and, in relation to plants or trees, gathering or cutting."

Cover Crop

A cover crop is a crop planted primarily to manage soil erosion, soil fertility, soil quality, water, weeds, pests, diseases, biodiversity and wildlife in an *agroecosystem* (Lu *et al.* 2000), an ecological system managed and largely shaped by humans across a range of intensities to produce food, feed, or fiber. Currently, not many countries are known for using the cover crop method.

Cover crops are of interest in sustainable agriculture as many of them improve the sustainability of agroecosystem attributes and may also indirectly improve qualities of neighboring natural eco-systems. Farmers choose to grow and manage specific cover crop types based on their own needs and goals, influenced by the biological, environmental, social, cultural, and economic factors of the food system in which farmers operate (Snapp *et al.* 2005). The farming practice of cover crops has been recognized as climate-smart agriculture by the White House.

Soil Erosion

Although cover crops can perform multiple functions in an agroecosystem simultaneously, they are often grown for the sole purpose of preventing soil erosion. Soil erosion is a process that can irreparably reduce the productive capacity of an agroecosystem. Dense cover crop stands physically slow down the velocity of rainfall before it contacts the soil surface, preventing soil splashing and erosive surface runoff (Romkens *et al.* 1990). Additionally, vast cover crop root networks help anchor the soil in place and increase soil porosity, creating suitable habitat networks for soil macrofauna (Tomlin *et al.* 1995).

Soil Fertility Management

One of the primary uses of cover crops is to increase soil fertility. These types of cover crops are referred to as "green manure." They are used to manage a range of soil macronutrients and micronutrients. Of the various nutrients, the impact that cover crops have on nitrogen management has received the most attention from researchers and farmers, because nitrogen is often the most limiting nutrient in crop production.

Often, green manure crops are grown for a specific period, and then plowed under before reaching full maturity in order to improve soil fertility and quality. Also the stalks left block the soil from being eroded.

Green manure crops are commonly leguminous, meaning they are part of the Fabaceae (pea) family.This family is unique in that all of the species in it set pods, such as bean, lentil, lupins and alfalfa. Leguminous cover crops are typically high in nitrogen and can often provide the required quantity of nitrogen for crop production. In conventional farming, this nitrogen is typically applied in chemical fertilizer form. This quality of cover crops is called fertilizer replacement value (Thiessen-Martens *et al.* 2005).

Another quality unique to leguminous cover crops is that they form symbiotic relationships with the rhizobial bacteria that reside in legume root nodules. Lupins is nodulated by the soil microorganism *Bradyrhizobium* sp. (Lupinus). Bradyrhizobia are encountered as microsymbionts in other leguminous crops (*Argyrolobium, Lotus, Ornithopus, Acacia, Lupinus*) of Mediterranean origin. These bacteria convert biologically unavailable atmospheric nitrogen gas (N_2) to biologically available ammonium (NH^+_4) through the process of biological nitrogen fixation.

Prior to the advent of the Haber-Bosch process, an energy-intensive method developed to carry out industrial nitrogen fixation and create chemical nitrogen fertilizer, most nitrogen introduced to ecosystems arose through biological nitrogen fixation (Galloway *et al.* 1995). Some scientists believe that widespread biological nitrogen fixation, achieved mainly through the use of cover crops, is the only alternative to industrial nitrogen fixation in the effort to maintain or increase future food production levels (Bohlool *et al.* 1992, Peoples and Craswell 1992, Giller and Cadisch 1995). Industrial nitrogen fixation has been criticized as an unsustainable source of nitrogen for food production due to its reliance on fossil fuel energy and the environmental impacts associated with chemical nitrogen fertilizer use in agriculture (Jensen and Hauggaard-Nielsen 2003). Such widespread environmental impacts include nitrogen fertilizer losses into waterways, which can lead to eutrophication (nutrient loading) and ensuing hypoxia (oxygen depletion) of large bodies of water.

An example of this lies in the Mississippi Valley Basin, where years of fertilizer nitrogen loading into the watershed from agricultural production have resulted in a hypoxic "dead zone" off the Gulf of Mexico the size of New Jersey (Rabalais *et al.* 2002). The ecological complexity of marine life in this zone has been diminishing as a consequence (CENR 2000).

As well as bringing nitrogen into agroecosystems through biological nitrogen fixation, types of cover crops known as "catch crops" are used to retain and recycle soil nitrogen already present. The catch crops take up surplus nitrogen remaining from fertilization of the previous crop, preventing it from being lost through leaching (Morgan *et al.* 1942), or gaseous denitrification or volatilization (Thorup-Kristensen *et al.* 2003).

Catch crops are typically fast-growing annual cereal species adapted to scavenge available nitrogen efficiently from the soil (Ditsch and Alley 1991). The nitrogen tied up in catch crop biomass is released back into the soil once the catch crop is incorporated as a green manure or otherwise begins to decompose.

An example of green manure use comes from Nigeria, where the cover crop *Mucuna pruriens* (velvet bean) has been found to increase the availability of phosphorus in soil after a farmer applies rock phosphate (Vanlauwe *et al.* 2000).

Soil Quality Management

Cover crops can also improve soil quality by increasing soil organic matter levels through the input of cover crop biomass over time. Increased soil organic matter enhances soil structure, as well as the water and nutrient holding and buffering capacity of soil (Patrick *et al.* 1957). It can also lead to increased soil carbon sequestration, which has been promoted as a strategy to help offset the rise in atmospheric carbon dioxide levels (Kuo *et al.* 1997, Sainju *et al.* 2002, Lal 2003).

Soil quality is managed to produce optimum circumstances for crops to flourish. The principal factors of soil quality are soil salination, pH, microorganism balance and the prevention of soil contamination.

Water Management

By reducing soil erosion, cover crops often also reduce both the rate and quantity of water that drains off the field, which would normally pose environmental risks to waterways and ecosystems downstream (Dabney *et al.* 2001). Cover crop biomass acts as a physical barrier between rainfall and the soil surface, allowing raindrops to steadily trickle down through the soil profile. Also, as stated above, cover crop root growth results in the formation of soil pores, which in addition to enhancing soil macrofauna habitat provides pathways for water to filter through the soil profile rather than draining off the field as surface flow. With increased water infiltration, the potential for soil water storage and the recharging of aquifers can be improved (Joyce *et al.* 2002).

Just before cover crops are killed (by such practices including mowing, tilling, discing, rolling, or herbicide application) they contain a large amount of moisture. When the cover crop is incorporated into the soil, or left on the soil surface, it often increases soil moisture. In agroecosystems where water for crop production is in short supply, cover crops can be used as a mulch to conserve water by shading and cooling the soil surface. This reduces evaporation of soil moisture. In other situations farmers try to dry the soil out as quickly as possible going into the planting season. Here prolonged soil moisture conservation can be problematic.

While cover crops can help to conserve water, in temperate regions (particularly in years with below average precipitation) they can draw down soil water supply in the spring, particularly if climatic growing conditions are good. In these cases, just before crop planting, farmers often face a tradeoff between the benefits of increased cover crop growth and the drawbacks of reduced soil moisture for cash crop production that season. C/N ratio is balanced with this application.

Weed Management

Thick cover crop stands often compete well with weeds during the cover crop growth period, and can prevent most germinated weed seeds from completing their life cycle and reproducing. If the cover crop is left on the soil surface rather than incorporated into the soil as a green manure after its growth is terminated, it can form a nearly impenetrable mat. This drastically reduces light transmittance to weed seeds, which in many cases reduces weed seed germination rates (Teasdale 1993). Furthermore, even when weed seeds germinate, they often run out of stored energy for growth before building the necessary structural capacity to break through the cover crop mulch layer. This is often termed the *cover crop smother effect* (Kobayashi *et al.* 2003).

Cover crop in South Dakota

Some cover crops suppress weeds both during growth and after death (Blackshaw *et al.* 2001). During growth these cover crops compete vigorously with weeds for available space, light, and nutrients, and after death they smother the next flush of weeds by forming a mulch layer on the soil surface. For example, Blackshaw *et al.* (2001) found that when using *Melilotus officinalis* (yellow sweetclover) as a cover crop in an improved fallow system (where a fallow period is intentionally improved by any number of different management practices, including the planting of cover crops), weed biomass only constituted between 1-12% of total standing biomass at the end of the cover crop growing season. Furthermore, after cover crop termination, the yellow sweetclover residues suppressed weeds to levels 75-97% lower than in fallow (no yellow sweetclover) systems.

In addition to competition-based or physical weed suppression, certain cover crops are known to suppress weeds through allelopathy (Creamer *et al.* 1996, Singh *et al.* 2003). This occurs when certain biochemical cover crop compounds are degraded that happen to be toxic to, or inhibit seed germination of, other plant species. Some well known examples of allelopathic cover crops are *Secale cereale* (rye), *Vicia villosa* (hairy vetch), *Trifolium pratense* (red clover), *Sorghum bicolor* (sorghum-sudangrass), and species in the Brassicaceae family, particularly mustards (Haramoto and Gallandt 2004). In one study, rye cover crop residues were found to have provided between 80% and 95% control of early season broadleaf weeds when used as a mulch during the production of different cash crops such as soybean, tobacco, corn, and sunflower (Nagabhushana *et al.* 2001).

In a recent study released by the Agricultural Research Service (ARS) scientists examined how rye seeding rates and planting patterns affected cover crop production. The results show that planting more pounds per acre of rye increased the cover crop's production as well as decreased the amount of weeds. The same was true when scientists tested seeding rates on legumes and oats; a higher density of seeds planted per acre decreased the amount of weeds and increased the yield of legume and oat production. The planting patterns, which consisted of either traditional rows or grid patterns, did not seem to make a significant impact on the cover crop's production or on the weed production in either cover crop. The ARS scientists concluded that increased seeding rates could be an effective method of weed control.

Disease Management

In the same way that allelopathic properties of cover crops can suppress weeds, they can also break disease cycles and reduce populations of bacterial and fungal diseases (Everts 2002), and parasitic nematodes (Potter *et al.* 1998, Vargas-Ayala *et al.* 2000). Species in the Brassicaceae family, such as mustards, have been widely shown to suppress fungal disease populations through the release of naturally occurring toxic chemicals during the degradation of glucosinolade compounds in their plant cell tissues (Lazzeri and Manici 2001).

Pest Management

Some cover crops are used as so-called "trap crops", to attract pests away from the crop of value and toward what the pest sees as a more favorable habitat (Shelton and Badenes-Perez 2006). Trap crop areas can be established within crops, within farms, or within landscapes. In many cases the trap crop is grown during the same season as the food crop being produced. The limited area occupied by these trap crops can be treated with a pesticide once pests are drawn to the trap in large enough numbers to reduce the pest populations. In some organic systems, farmers drive over the trap crop with a large vacuum-based implement to physically pull the pests off the plants and out of the field (Kuepper and Thomas 2002). This system has been recommended for use to help control the lygus bugs in organic strawberry production (Zalom *et al.* 2001). Another example of trap crops are nematode resistance White mustard (*Sinapis alba*) and Radish (*Raphanus sativus*). They can be grown after a main (cereal) crop and trap nematodes, for example the beet cyst nematode and Columbian root knot nematode. When grown, nematodes hatch and are attracted to the roots. After entering the roots they cannot reproduce in the root due to a hypersensitive resistance reaction of the plant. Hence the nematode population is greatly reduced, by 70-99%, depending on species and cultivation time.

Other cover crops are used to attract natural predators of pests by providing elements of their habitat. This is a form of biological control known as habitat augmentation, but achieved with the use of cover crops (Bugg and Waddington 1994). Findings on the relationship between cover crop presence and predator/pest population dynamics have been mixed, pointing toward the need for detailed information on specific cover crop types and management practices to best complement a given integrated pest management strategy. For example, the predator mite *Euseius tularensis* (Congdon) is known to help control the pest citrus thrips in Central California citrus orchards. Researchers found that the planting of several different leguminous cover crops (such as bell bean, woollypod vetch, New Zealand white clover, and Austrian winter pea) provided sufficient pollen as a feeding source to cause a seasonal increase in *E. tularensis* populations, which with good timing could potentially introduce enough predatory pressure to reduce pest populations of citrus thrips (Grafton-Cardwell *et al.* 1999).

Diversity and Wildlife

Although cover crops are normally used to serve one of the above discussed purposes, they often simultaneously improve farm habitat for wildlife. The use of cover crops adds at least one more dimension of plant diversity to a cash crop rotation. Since the cover crop is typically not a crop of value, its management is usually less intensive, providing a window of "soft" human influence on the farm. This relatively "hands-off" management, combined with the increased on-farm heterogeneity created by the establishment of cover crops, increases the likelihood that a more complex trophic structure will develop to support a higher level of wildlife diversity (Freemark and Kirk 2001).

In one study, researchers compared arthropod and songbird species composition and field use between conventionally and cover cropped cotton fields in the Southern United States. The cover cropped cotton fields were planted to clover, which was left to grow in between cotton rows throughout the early cotton growing season (stripcover cropping). During the migration and breeding season, they found that songbird densities were 7–20 times higher in the cotton fields with integrated clover cover crop than in the conventional cotton fields. Arthropod abundance and biomass was also higher in the clover cover cropped fields throughout much of the songbird breeding season, which was attributed to an increased supply of flower nectar from the clover. The clover cover crop enhanced songbird habitat by providing cover and nesting sites, and an increased food source from higher arthropod populations (Cederbaum *et al.* 2004).

Intercropping

Intercropping is a multiple cropping practice involving growing two or more crops in proximity. The most common goal of intercropping is to produce a greater yield on a given piece of land by making use of resources that would otherwise not be utilized by a single crop. Careful planning is required, taking into account the soil, climate, crops, and varieties. It is particularly important not to have crops competing with each other for physical space, nutrients, water, or sunlight. Examples of intercropping strategies are planting a deep-rooted crop with a shallow-rooted crop, or planting a tall crop with a shorter crop that requires partial shade. Inga alley cropping has been proposed as an alternative to the ecological destruction of slash-and-burn farming.

When crops are carefully selected, other agronomic benefits are also achieved. Lodging-prone plants, those that are prone to tip over in wind or heavy rain, may be given structural support by their companion crop. Creepers can also benefit from structural support. Some plants are used to suppress weeds or provide nutrients. Delicate or light-sensitive plants may be given shade or protection, or otherwise wasted space can be utilized. An example is the tropical multi-tier system where coconut occupies the upper tier, banana the middle tier, and pineapple, ginger, or leguminous fodder, medicinal or aromatic plants occupy the lowest tier.

Intercropping of compatible plants also encourages biodiversity, by providing a habitat for a variety of insects and soil organisms that would not be present in a single-crop environment. This in turn can help limit outbreaks of crop pests by increasing predator biodiversity. Additionally, reducing the homogeneity of the crop increases the barriers against biological dispersal of pest organisms through the crop.

The degree of spatial and temporal overlap in the two crops can vary somewhat, but both requirements must be met for a cropping system to be an intercrop. Numerous types of intercropping, all of which vary the temporal and spatial mixture to some degree, have been identified. These are some of the more significant types:

- Mixed intercropping, as the name implies, is the most basic form in which the component crops are totally mixed in the available space.

- Row cropping involves the component crops arranged in alternate rows. Variations include alley cropping, where crops are grown in between rows of trees, and strip cropping, where multiple rows, or a strip, of one crop are alternated with multiple rows of another crop. A new version of this is to intercrop rows of solar photovoltaic modules with agriculture crops. This practice is called agrivoltaics.

- Temporal intercropping uses the practice of sowing a fast-growing crop with a slow-growing crop, so that the fast-growing crop is harvested before the slow-growing crop starts to mature.

- Further temporal separation is found in relay cropping, where the second crop is sown during the growth, often near the onset of reproductive development or fruiting, of the first crop, so that the first crop is harvested to make room for the full development of the second.

Chili pepper intercropped with coffee in Colombia's southwestern Cauca Department

Coconut and *Tagetes erecta*, a multilayer cropping in India

Crop Rotation

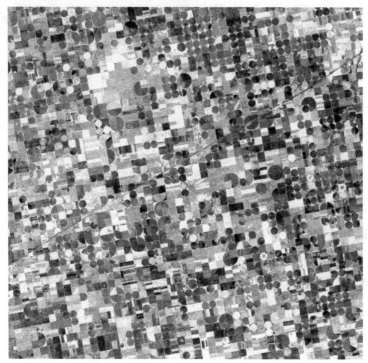

Satellite image of circular crop fields in Kansas in late June 2001.

Healthy, growing crops are green. Corn would be growing into leafy stalks by then. Sorghum, which resembles corn, grows more slowly and would be much smaller and therefore, (possibly) paler. Wheat is a brilliant yellow as harvest occurs in June. Fields of brown have been recently harvested and plowed under or lie fallow for the year.

Effects of crop rotation and monoculture at the Swojec Experimental Farm, Wroclaw University of Environmental and Life Sciences. In the front field, the "Norfolk" crop rotation sequence (potatoes, oats, peas, rye) is being applied; in the back field, rye has been grown for 45 years in a row.

Crop rotation is the practice of growing a series of dissimilar or different types of crops in the same area in sequenced seasons.It is done so that the soil of farms is not used to only one type of nutrient. It helps in reducing soil erosion and increases soil fertility and crop yield.

Growing the same crop in the same place for many years in a row disproportionately depletes the soil of certain nutrients. With rotation, a crop that leaches the soil of one kind of nutrient is followed during the next growing season by a dissimilar crop that returns that nutrient to the soil or draws a different ratio of nutrients. In addition, crop rotation mitigates the buildup of pathogens and pests that often occurs when one species is continuously cropped, and can also improve soil structure and fertility by increasing biomass from varied root structures.

Crop rotation is used in both conventional and organic farming systems.

History

It has long been recognized that suitable rotations – such as planting spring crops for livestock in place of grains for human consumption – make it possible to restore or to maintain a productive soil. Middle Eastern farmers practiced crop rotation in 6000 BC without understanding the chemistry, alternately planting legumes and cereals. In the Bible chapter of Leviticus 25, God instructs the Israelites to observe a 'Sabbath of the Land'. Every seventh year they would not till, prune or even control insects. The Roman writer, Cato the Elder, recommended that farmers "save carefully goat, sheep, cattle, and all other dung". In Europe, since the times of Charlemagne, there was a transition from a two-field crop rotation to a three-field crop rotation. Under a two-field rotation, half the land was planted in a year, while the other half lay fallow. Then, in the next year, the two fields were reversed.

From the end of the Middle Ages until the 20th century, three-year rotation was practiced by farmers in Europe. Under three-field rotation, the land was divided into three parts. One section was planted in the autumn with rye or winter wheat, followed by spring oats or barley, or other crops such as peas, lentils, or beans and the third field was left fallow. The three fields were rotated in this manner so that every three years, a field would rest and be fallow. Under the two-field system, if one has a total of 600 acres (2.4 km²) of fertile land, one would only plant 300 acres. Under the new three-field rotation system, one would plant (and therefore harvest) 400 acres. But, the additional crops had a more significant effect than mere productivity. Since the spring crops were mostly legumes, they increased the overall nutrition of the people of Northern Europe.

A four-field rotation was pioneered by farmers, namely in the region Waasland in the early 16th century and popularised by the British agriculturist Charles Townshend in the 18th century. The system (wheat, turnips, barley and clover), opened up a fodder crop and grazing crop allowing livestock to be bred year-round. The four-field crop rotation was a key development in the British Agricultural Revolution.

George Washington Carver studied crop rotation methods in the United States, teaching southern farmers to rotate soil-depleting crops like cotton with soil-enriching crops like peanuts and peas.

In the Green Revolution, the traditional practice of crop rotation gave way in some parts of the world to the practice of supplementing the chemical inputs to the soil through top dressing with fertilizers, e.g. adding ammonium nitrate or urea and restoring soil pH with lime in the search for increased yields, preparing soil for specialist crops, and seeking to reduce waste and inefficiency by simplifying planting and harvesting.

Crop Choice

A preliminary assessment of crop interrelationships can be found in how each crop: (1) contributes to soil organic matter (SOM) content, (2) provides for pest management, (3) manages deficient or excess nutrients, and (4) how it contributes to or controls for soil erosion.

Crop choice is often a related to the goal the farmer is looking to achieve with the rotation, which could be weed management, increasing available nitrogen in the soil, controlling for erosion, or increasing soil structure and biomass, to name a few. When discussing crop rotations, crops are classified in different ways depending on what quality is being assessed: by family, by nutrient needs/benefits, and/or by profitability (i.e. cash crop versus cover crop). For example, giving adequate attention to plant family is essential to mitigating pests and pathogens. However, many farmers have success managing rotations by planning sequencing and cover crops around desirable cash crops. The following is a simplified classification based on crop quality and purpose.

Row Crops

Many crops which are critical for the market, like vegetables, are row crops (that is, grown in tight rows). While often the most profitable for farmers, these crops are more taxing on the soil Row crops typically have low biomass and shallow roots: this means the plant contributes low residue to the surrounding soil and has limited effects on structure. With much of the soil around the plant is exposed to disruption by rainfall and traffic, fields with row crops experience faster break down of organic matter by microbes, leaving fewer nutrients for future plants.

In short, while these crops may be profitable for the farm, they are nutrient depleting. Crop rotation practices exist to strike a balance between short-term profitability and long-term productivity.

Legumes

A great advantage of crop rotation comes from the interrelationship of nitrogen fixing-crops with nitrogen demanding crops. Legumes, like alfalfa and clover, collect available nitrogen from the soil in nodules on their root structure. When the plant is harvested, the biomass of uncollected roots breaks down, making the stored nitrogen available to future crops. Legumes are also a valued green manure: a crop that collects nutrients and fixes them at soil depths accessible to future crops.

In addition, legumes have heavy tap roots that burrow deep into the ground, lifting soil for better tilth and absorption of water.

Grasses and Cereals

Cereal and grasses are frequent cover crops because of the many advantages they supply to soil quality and structure. The dense and far-reaching root systems give ample structure to surrounding soil and provide significant biomass for soil organic matter.

Grasses and cereals are key in weed management as they compete with undesired plants for soil space and nutrients.

Green Manure

Green manure is a crop that is mixed into the soil. Both nitrogen-fixing legumes and nutrient scavengers, like grasses, can be used as green manure. Green manure of legumes is an excellent source of nitrogen, especially for organic systems, however, legume biomass doesn't contribute to lasting soil organic matter like grasses do.

Planning a Rotation

There are numerous factors that must be taken into consideration when planning a crop rotation. Planning an effective rotation requires weighing fixed and fluctuating production circumstances, including, but not limited to: market, farm size, labor supply, climate, soil type, growing practices, etc. Moreover, a crop rotation must consider in what condition one crop will leave the soil for the succeeding crop and how one crop can be seeded with another crop. For example, a nitrogen-fixing crop, like a legume, should always proceed a nitrogen depleting one; similarly, a low residue crop (i.e. a crop with low biomass) should be offset with a high biomass cover crop, like a mixture of grasses and legumes.

There is no limit to the number of crops that can be used in a rotation, or the amount of time a rotation takes to complete. Decisions about rotations are made years prior, seasons prior, or even at the very last minute when an opportunity to increase profits or soil quality presents itself. In short, there is no singular formula for rotation, but many considerations to take into account.

Implementation

Crop rotation systems may be enriched by the influences of other practices such as the addition of livestock and manure, intercropping or multiple cropping, and organic management low in pesticides and synthetic fertilizers.

Incorporation of Livestock

Introducing livestock makes the most efficient use of critical sod and cover crops; livestock (through manure) are able to distribute the nutrients in these crops throughout the soil rather than removing nutrients from the farm through the sale of hay. In systems where use of farm livestock would violate reservations growers or consumers may have about animal exploitation, efforts are made to surrogate this input through livestock in the soil, namely worms and microorganisms.

In Sub-Saharan Africa, as animal husbandry becomes less of a nomadic practice many herders have begun integrating crop production into their practice. This is known as mixed farming, or the practice of crop cultivation with the incorporation of raising cattle, sheep and/or goats by the same economic entity, is increasingly common. This interaction between the animal, the land and the crops are being done on a small scale all across this region. Crop residues provide animal feed, while the animals provide manure for replenishing crop nutrients and draft power. Both processes are extremely important in this region of the world as it is expensive and logistically unfeasible to transport in synthetic fertilizers and large-scale machinery. As an additional benefit, the cattle, sheep and/or goat provide milk and can act as a cash crop in the times of economic hardship.

Organic Farming

Crop rotation is a required practice in order for a farm to receive organic certification in the United States. The "Crop Rotation Practice Standard" for the National Organic Program under the U.S. Code of Federal Regulations, section §205.205, states that:

Farmers are required to implement a crop rotation that maintains or builds soil organic matter, works to control pests, manages and conserves nutrients, and protects against erosion. Producers of perennial crops that aren't rotated may utilize other practices, such as cover crops, to maintain soil health.

In addition to lowering the need for inputs by controlling for pests and weeds and increasing available nutrients, crop rotation helps organic growers increase the amount of biodiversity on their farms. Biodiversity is also a requirement of organic certification, however, there are no rules in place to regulate or reinforce this standard. Increasing the biodiversity of crops has beneficial effects on the surrounding ecosystem and can host a greater diversity of fauna, insects, and beneficial microorganism in the soil.< Some studies point to increased nutrient availability from crop rotation under organic systems compared to conventional practices as organic practices are less likely to inhibit of beneficial microbes in soil organic matter.

While multiple cropping and intercropping benefit from many of the same principals as crop rotation, they do not satisfy the requirement under the NOP.

Intercropping

Multiple cropping systems, such as intercropping or companion planting, offer more diversity and complexity within the same season or rotation, for example the three sisters. An example of companion planting is the inter-planting of corn with pole beans and vining squash or pumpkins. In this system, the beans provide nitrogen; the corn provides support for the beans and a "screen" against squash vine borer; the vining squash provides a weed suppressive canopy and discourages corn-hungry raccoons.

Double-cropping is common where two crops, typically of different species, are grown sequentially in the same growing season, or where one crop (e.g. vegetable) is grown continuously with a cover crop (e.g. wheat). This is advantageous for small farms, who often cannot afford to leave cover crops to replenish the soil for extended periods of time, as larger farms can. When multiple cropping is implemented on small farms, these systems can maximize benefits of crop rotation on available land resources.

Benefits

Agronomists describe the benefits to yield in rotated crops as "The Rotation Effect". There are many found benefits of rotation systems: however, there is no specific scientific basis for the sometimes 10-25% yield increase in a crop grown in rotation versus monoculture. The factors related to the increase are simply described as alleviation of the negative factors of monoculture cropping systems. Explanations due to improved nutrition; pest, pathogen, and weed stress reduction; and improved soil structure have been found in some cases to be correlated, but causation has not been determined for the majority of cropping systems.

Other benefits of rotation cropping systems include production cost advantages. Overall financial risks are more widely distributed over more diverse production of crops and/or livestock. Less reliance is placed on purchased inputs and over time crops can maintain production goals with fewer inputs. This in tandem with greater short and long term yields makes rotation a powerful tool for improving agricultural systems.

Soil Organic Matter

The use of different species in rotation allows for increased soil organic matter (SOM), greater soil structure, and improvement of the chemical and biological soil environment for crops. With more SOM, water infiltration and retention improves, providing increased drought tolerance and decreased erosion.

Soil organic matter is a mix of decaying material from biomass with active microorganisms. Crop rotation, by nature, increases exposure to biomass from sod, green manure, and a various other plant debris. The reduced need for intensive tillage under crop rotation allows biomass aggregation to lead to greater nutrient retention and utilization, decreasing the need for added nutrients. With tillage, disruption and oxidation of soil creates a less conducive environment for diversity and proliferation of microorganisms in the soil. These microorganisms are what make nutrients available to plants. So, where "active" soil organic matter is a key to productive soil, soil with low microbial activity provides significantly fewer nutrients to plants; this is true even though the quantity of biomass left in the soil may be the same.

Soil microorganisms also decrease pathogen and pest activity through competition. In addition, plants produce root exudates and other chemicals which manipulate their soil environment as well as their weed environment. Thus rotation allows increased yields from nutrient availability but also alleviation of allelopathy and competitive weed environments.

Carbon Sequestration

Studies have shown that crop rotations greatly increase soil organic carbon (SOC) content, the main constituent of soil organic matter. Carbon, along with hydrogen and oxygen, is a macronutrient for plants. Highly diverse rotations spanning long periods of time have shown to be even more effective in increasing SOC, while soil disturbances (e.g. from tillage) are responsible for exponential decline in SOC levels. In Brazil, conservation to no-till methods combined with intensive crop rotations has been shown an SOC sequestration rate of 0.41 tonnes per hectare per year.

In addition to enhancing crop productivity, sequestration of atmospheric carbon has great implications in reducing rates of climate change by removing carbon dioxide from the air.

Nitrogen Fixing

Rotating crops adds nutrients to the soil. Legumes, plants of the family Fabaceae, for instance, have nodules on their roots which contain nitrogen-fixing bacteria called rhizobia. It therefore makes good sense agriculturally to alternate them with cereals (family Poaceae) and other plants that require nitrates.

Pathogen and Pest Control

Crop rotation is also used to control pests and diseases that can become established in the soil over time. The changing of crops in a sequence decreases the population level of pests by (1) interrupting pest life cycles and (2) interrupting pest habitat. Plants within the same taxonomic family tend to have similar pests and pathogens. By regularly changing crops and keeping the soil occupied by cover crops instead of lying fallow, pest cycles can be broken or limited, especially cycles that benefit from overwintering in residue. For example, root-knot nematode is a serious problem for some plants in warm climates and sandy soils, where it slowly builds up to high levels in the soil, and can severely damage plant productivity by cutting off circulation from the plant roots. Growing a crop that is not a host for root-knot nematode for one season greatly reduces the level of the nematode in the soil, thus making it possible to grow a susceptible crop the following season without needing soil fumigation.

This principle is of particular use in organic farming, where pest control must be achieved without synthetic pesticides.

Weed Management

Integrating certain crops, especially cover crops, into crop rotations is of particular value to weed management. These crops crowd out weed through competition. In addition, the sod and compost from cover crops and green manure slows the growth of what weeds are still able to make it through the soil, giving the crops further competitive advantage. By removing slowing the growth and proliferation of weeds while cover crops are cultivated, farmers greatly reduce the presence of weeds for future crops, including shallow rooted and row crops, which are less resistant to weeds. Cover crops are, therefore, considered conservation crops because they protect otherwise fallow land from becoming overrun with weeds.

This system has advantages over other common practices for weeds management, such as tillage.

Tillage is meant to inhibit growth of weeds by overturning the soil; however, this has a countering effect of exposing weed seeds that may have gotten buried and burying valuable crop seeds. Under crop rotation, the number of viable seeds in the soil is reduced through the reduction of the weed population.

Preventing Soil Erosion

Crop rotation can significantly reduce the amount of soil lost from erosion by water. In areas that are highly susceptible to erosion, farm management practices such as zero and reduced tillage can be supplemented with specific crop rotation methods to reduce raindrop impact, sediment detachment, sediment transport, surface runoff, and soil loss.

Protection against soil loss is maximized with rotation methods that leave the greatest mass of crop stubble (plant residue left after harvest) on top of the soil. Stubble cover in contact with the soil minimizes erosion from water by reducing overland flow velocity, stream power, and thus the ability of the water to detach and transport sediment. Soil Erosion and Cill prevent the disruption and detachment of soil aggregates that cause macropores to block, infiltration to decline, and runoff to increase. This significantly improves the resilience of soils when subjected to periods of erosion and stress.

The effect of crop rotation on erosion control varies by climate. In regions under relatively consistent climate conditions, where annual rainfall and temperature levels are assumed, rigid crop rotations can produce sufficient plant growth and soil cover. In regions where climate conditions are less predictable, and unexpected periods of rain and drought may occur, a more flexible approach for soil cover by crop rotation is necessary. An opportunity cropping system promotes adequate soil cover under these erratic climate conditions. In an opportunity cropping system, crops are grown when soil water is adequate and there is a reliable sowing window. This form of cropping system is likely to produce better soil cover than a rigid crop rotation because crops are only sown under optimal conditions, whereas rigid systems are not necessarily sown in the best conditions available.

Crop rotations also affect the timing and length of when a field is subject to fallow. This is very important because depending on a particular region's climate, a field could be the most vulnerable to erosion when it is under fallow. Efficient fallow management is an essential part of reducing erosion in a crop rotation system. Zero tillage is a fundamental management practice that promotes crop stubble retention under longer unplanned fallows when crops cannot be planted. Such management practices that succeed in retaining suitable soil cover in areas under fallow will ultimately reduce soil loss.

Biodiversity

Increasing the biodiversity of crops has beneficial effects on the surrounding ecosystem and can host a greater diversity of fauna, insects, and beneficial microorganisms in the soil. Some studies point to increased nutrient availability from crop rotation under organic systems compared to conventional practices as organic practices are less likely to inhibit of beneficial microbes in soil organic matter, such as arbuscular mycorrhizae, which increase nutrient uptake in plants. Increasing biodiversity also increases the resilience of agro-ecological systems.

Farm Productivity

Crop rotation contributes to increased yields through improved soil nutrition. By requiring planting and harvesting of different crops at different times, more land can be farmed with the same amount of machinery and labour.

Risk Management

Different crops can reduce the risks of adverse weather for the individual farmer.

Challenges

While crop rotation requires a great deal of planning, crop choice must respond to a number of fixed conditions (soil type, topography, climate, and irrigation) in addition to conditions that may change dramatically from year to the next (weather, market, labor supply). In this way, it is unwise to plan to crops years in advance. Improper implementation of a crop rotation plan may lead to imbalances in the soil nutrient composition or a buildup of pathogens affecting a critical crop. The consequences of faulty rotation may take years to become apparent even to experienced soil scientists and can take just as long to correct.

Many challenges exist within the practices associated with crop rotation. For example, green manure from legumes can lead to an invasion of snails or slugs and the decay from green manure can occasionally suppress the growth of other crops.

Hydroculture

A NASA researcher checking hydroponically cultivated onions, with Bibb lettuce to his left and radishes to the right.

Hydroculture is the growing of plants in a soilless medium, or an aquatic based environment. Plant nutrients are distributed via water.

The word "hydro" derives its name from hudōr meaning water, hence hydroculture = water culture. Hydroculture is aquatic horticulture.

Techniques

In basic hydroculture or passive hydroponics, water and nutrients are distributed through capillary action. In hydroponics-like hydroculture, water and nutrients are distributed by some form of pumping mechanism.

Substrates

The roots might be anchored in clay aggregate such as the trademarks LECA and Hydroton.

Advantages include ease of maintenance as watering and feeding involve just topping up the reservoir of growing solution. Certain types of hydroponic media are resistant to some types of soil-borne insects.

Water plant cultivated crocus

Expanded clay pellets

Plant Breeding

Plant breeding is the art and science of changing the traits of plants in order to produce desired characteristics. Plant breeding can be accomplished through many different techniques ranging from simply selecting plants with desirable characteristics for propagation, to more complex molecular techniques.

The Yecoro wheat (right) cultivar is sensitive to salinity, plants resulting from a hybrid cross with cultivar W4910 (left) show greater tolerance to high salinity

Plant breeding has been practiced for thousands of years, since near the beginning of human civilization. It is practiced worldwide by individuals such as gardeners and farmers, or by professional plant breeders employed by organizations such as government institutions, universities, crop-specific industry associations or research centers.

International development nation agencies believe that breeding new crops is important for ensuring food security by developing new varieties that are higher-yielding, disease resistant, drought-resistant or regionally adapted to different environments and growing conditions.

History

Plant breeding started with sedentary agriculture and particularly the domestication of the first agricultural plants, a practice which is estimated to date back 9,000 to 11,000 years. Initially early farmers simply selected food plants with particular desirable characteristics, and employed these as progenitors for subsequent generations, resulting in an accumulation of valuable traits over time.

Gregor Mendel's experiments with plant hybridization led to his establishing laws of inheritance. Once this work became well known, it formed the basis of the new science of genetics, which stimulated research by many plant scientists dedicated to improving crop production through plant breeding.

Modern plant breeding is applied genetics, but its scientific basis is broader, covering molecular biology, cytology, systematics, physiology, pathology, entomology, chemistry, and statistics (biometrics). It has also developed its own technology.

Classical Plant Breeding

One major technique of plant breeding is selection, the process of selectively propagating plants with desirable characteristics and eliminating or "culling" those with less desirable characteristics.

Another technique is the deliberate interbreeding (crossing) of closely or distantly related individuals to produce new crop varieties or lines with desirable properties. Plants are crossbred to introduce traits/genes from one variety or line into a new genetic background. For example, a mildew-resistant pea may be crossed with a high-yielding but susceptible pea, the goal of the cross being to introduce mildew resistance without losing the high-yield characteristics. Progeny from the cross would then be crossed with the high-yielding parent to ensure that the progeny were most like the high-yielding parent, (backcrossing). The progeny from that cross would then be tested for yield (selection, as described above) and mildew resistance and high-yielding resistant plants would be further developed. Plants may also be crossed with themselves to produce inbred varieties for breeding. Pollinators may be excluded through the use of pollination bags.

Classical breeding relies largely on homologous recombination between chromosomes to generate genetic diversity. The classical plant breeder may also make use of a number of *in vitro* techniques such as protoplast fusion, embryo rescue or mutagenesis to generate diversity and produce hybrid plants that would not exist in nature.

Traits that breeders have tried to incorporate into crop plants include:

1. Improved quality, such as increased nutrition, improved flavor, or greater beauty

2. Increased yield of the crop

3. Increased tolerance of environmental pressures (salinity, extreme temperature, drought)

4. Resistance to viruses, fungi and bacteria

5. Increased tolerance to insect pests

6. Increased tolerance of herbicides

7. Longer storage period for the harvested crop

Before World War II

Garton's catalogue from 1902

Successful commercial plant breeding concerns were founded from the late 19th century. Gartons Agricultural Plant Breeders in England was established in the 1890s by John Garton, who was one of the first to commercialize new varieties of agricultural crops created through cross-pollination. The firm's first introduction was Abundance Oat, one of the first agricultural grain varieties bred from a *controlled* cross, introduced to commerce in 1892.

In the early 20th century, plant breeders realized that Mendel's findings on the non-random nature of inheritance could be applied to seedling populations produced through deliberate pollinations to predict the frequencies of different types. Wheat hybrids were bred to increase the crop production of Italy during the so-called "Battle for Grain" (1925–1940). Heterosis was explained by George Harrison Shull. It describes the tendency of the progeny of a specific cross to outperform both parents. The detection of the usefulness of heterosis for plant breeding has led to the development of inbred lines that reveal a heterotic yield advantage when they are crossed. Maize was the first species where heterosis was widely used to produce hybrids.

Statistical methods were also developed to analyze gene action and distinguish heritable variation from variation caused by environment. In 1933 another important breeding technique, cytoplasmic male sterility (CMS), developed in maize, was described by Marcus Morton Rhoades. CMS is a maternally inherited trait that makes the plant produce sterile pollen. This enables the production of hybrids without the need for labor-intensive detasseling.

These early breeding techniques resulted in large yield increase in the United States in the early 20th century. Similar yield increases were not produced elsewhere until after World War II, the Green Revolution increased crop production in the developing world in the 1960s.

After World War II

In vitro-culture of Vitis (grapevine), Geisenheim Grape Breeding Institute

Following World War II a number of techniques were developed that allowed plant breeders to hybridize distantly related species, and artificially induce genetic diversity.

When distantly related species are crossed, plant breeders make use of a number of plant tissue culture techniques to produce progeny from otherwise fruitless mating. Interspecific and intergeneric hybrids are produced from a cross of related species or genera that do not normally sexually reproduce with each other. These crosses are referred to as *Wide crosses*. For example, the cereal triticale is a wheat and rye hybrid. The cells in the plants derived from the first generation created from the cross contained an uneven number of chromosomes and as result was sterile. The cell division inhibitor colchicine was used to double the number of chromosomes in the cell and thus allow the production of a fertile line.

Failure to produce a hybrid may be due to pre- or post-fertilization incompatibility. If fertilization is possible between two species or genera, the hybrid embryo may abort before maturation. If this does occur the embryo resulting from an interspecific or intergeneric cross can sometimes be rescued and cultured to produce a whole plant. Such a method is referred to as Embryo Rescue. This technique has been used to produce new rice for Africa, an interspecific cross of Asian rice (*Oryza sativa*) and African rice (*Oryza glaberrima*).

Hybrids may also be produced by a technique called protoplast fusion. In this case protoplasts are fused, usually in an electric field. Viable recombinants can be regenerated in culture.

Chemical mutagens like EMS and DMS, radiation and transposons are used to generate mutants with desirable traits to be bred with other cultivars - a process known as *Mutation Breeding*. Classical plant breeders also generate genetic diversity within a species by exploiting a process called somaclonal variation, which occurs in plants produced from tissue culture, particularly plants derived from callus. Induced polyploidy, and the addition or removal of chromosomes using a technique called chromosome engineering may also be used.

When a desirable trait has been bred into a species, a number of crosses to the favored parent are made to make the new plant as similar to the favored parent as possible. Returning to the example of the mildew resistant pea being crossed with a high-yielding but susceptible pea, to make the mildew resistant progeny of the cross most like the high-yielding parent, the progeny will be crossed back to that parent for several generations. This process removes most of the genetic contribution of the mildew resistant parent. Classical breeding is therefore a cyclical process.

With classical breeding techniques, the breeder does not know exactly what genes have been introduced to the new cultivars. Some scientists therefore argue that plants produced by classical breeding methods should undergo the same safety testing regime as genetically modified plants. There have been instances where plants bred using classical techniques have been unsuitable for human consumption, for example the poison solanine was unintentionally increased to unacceptable levels in certain varieties of potato through plant breeding. New potato varieties are often screened for solanine levels before reaching the marketplace.

Modern Plant Breeding

Modern plant breeding may use techniques of molecular biology to select, or in the case of genetic modification, to insert, desirable traits into plants. Application of biotechnology or molecular biology is also known as molecular breeding.

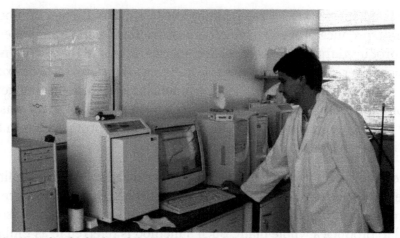

Modern facilities in molecular biology have converted classical plant breeding to molecular plant breeding

Steps of Plant Breeding

The following are the major activities of plant breeding:

- Collection of variation

- Selection

- Evaluation

- Release

- Multiplication

- Distribution of the new variety

Marker Assisted Selection

Sometimes many different genes can influence a desirable trait in plant breeding. The use of tools such as molecular markers or DNA fingerprinting can map thousands of genes. This allows plant breeders to screen large populations of plants for those that possess the trait of interest. The screening is based on the presence or absence of a certain gene as determined by laboratory procedures, rather than on the visual identification of the expressed trait in the plant.

Reverse Breeding and Doubled Haploids (DH)

A method for efficiently producing homozygous plants from a heterozygous starting plant, which has all desirable traits. This starting plant is induced to produce doubled haploid from haploid cells, and later on creating homozygous/doubled haploid plants from those cells. While in natural offspring genetic recombination occurs and traits can be unlinked from each other, in doubled haploid cells and in the resulting DH plants recombination is no longer an issue. There, a recombination between two corresponding chromosomes does not lead to un-linkage of alleles or traits, since it just leads to recombination with its identical copy. Thus, traits on one chromosome stay linked. Selecting those offspring having the desired set of chromosomes and crossing them will result in a final F1 hybrid plant, having exactly the same set of chromosomes, genes and traits as the starting hybrid plant. The homozygous parental lines can reconstitute the original heterozygous plant by crossing, if desired even in a large quantity. An individual heterozygous plant can be converted into a heterozygous variety (F1 hybrid) without the necessity of vegetative propagation but as the result of the cross of two homozygous/doubled haploid lines derived from the originally selected plant.

Genetic Modification

Genetic modification of plants is achieved by adding a specific gene or genes to a plant, or by knocking down a gene with RNAi, to produce a desirable phenotype. The plants resulting from adding a gene are often referred to as transgenic plants. If for genetic modification genes of the species or of a crossable plant are used under control of their native promoter, then they are called cisgenic plants. Sometimes genetic modification can produce a plant with the desired trait or traits faster than classical breeding because the majority of the plant's genome is not altered.

To genetically modify a plant, a genetic construct must be designed so that the gene to be added or removed will be expressed by the plant. To do this, a promoter to drive transcription and a termination sequence to stop transcription of the new gene, and the gene or genes of interest must be introduced to the plant. A marker for the selection of transformed plants is also included. In the laboratory, antibiotic resistance is a commonly used marker: Plants that have been successfully transformed will grow on media containing antibiotics; plants that have not been transformed will die. In some instances markers for selection are removed by backcrossing with the parent plant prior to commercial release.

The construct can be inserted in the plant genome by genetic recombination using the bacteria *Agrobacterium tumefaciens* or *A. rhizogenes*, or by direct methods like the gene gun or micro-injection. Using plant viruses to insert genetic constructs into plants is also a possibility, but the technique is limited by the host range of the virus. For example, Cauliflower mosaic virus (CaMV) only infects cauliflower and related species. Another limitation of viral vectors is that the virus is not usually passed on the progeny, so every plant has to be inoculated.

The majority of commercially released transgenic plants are currently limited to plants that have introduced resistance to insect pests and herbicides. Insect resistance is achieved through incorporation of a gene from *Bacillus thuringiensis* (Bt) that encodes a protein that is toxic to some insects. For example, the cotton bollworm, a common cotton pest, feeds on Bt cotton it will ingest the toxin and die. Herbicides usually work by binding to certain plant enzymes and inhibiting their action. The enzymes that the herbicide inhibits are known as the herbicides *target site*. Herbicide resistance can be engineered into crops by expressing a version of *target site* protein that is not inhibited by the herbicide. This is the method used to produce glyphosate resistant crop plants.

Genetic modification of plants that can produce pharmaceuticals (and industrial chemicals), sometimes called *pharming*, is a rather radical new area of plant breeding.

Issues and Concerns

Modern plant breeding, whether classical or through genetic engineering, comes with issues of concern, particularly with regard to food crops. The question of whether breeding can have a negative effect on nutritional value is central in this respect. Although relatively little direct research in this area has been done, there are scientific indications that, by favoring certain aspects of a plant's development, other aspects may be retarded. A study published in the *Journal of the American College of Nutrition* in 2004, entitled *Changes in USDA Food Composition Data for 43 Garden Crops, 1950 to 1999*, compared nutritional analysis of vegetables done in 1950 and in 1999, and found substantial decreases in six of 13 nutrients measured, including 6% of protein and 38% of riboflavin. Reductions in calcium, phosphorus, iron and ascorbic acid were also found. The study, conducted at the Biochemical Institute, University of Texas at Austin, concluded in summary: *"We suggest that any real declines are generally most easily explained by changes in cultivated varieties between 1950 and 1999, in which there may be trade-offs between yield and nutrient content."*

The debate surrounding genetically modified food during the 1990s peaked in 1999 in terms of media coverage and risk perception, and continues today - for example, *"Germany has thrown its weight behind a growing European mutiny over genetically modified crops by banning the planting of a widely grown pest-resistant corn variety."* The debate encompasses the ecological impact of genetically modified plants, the safety of genetically modified food and concepts used for safety evaluation like substantial equivalence. Such concerns are not new to plant breeding. Most countries have regulatory processes in place to help ensure that new crop varieties entering the marketplace are both safe and meet farmers' needs. Examples include variety registration, seed schemes, regulatory authorizations for GM plants, etc.

Plant breeders' rights is also a major and controversial issue. Today, production of new varieties is dominated by commercial plant breeders, who seek to protect their work and collect royalties through national and international agreements based in intellectual property rights. The range of related issues is complex. In the simplest terms, critics of the increasingly restrictive regulations argue that, through a combination of technical and economic pressures, commercial breeders are reducing biodiversity and significantly constraining individuals (such as farmers) from developing and trading seed on a regional level. Efforts to strengthen breeders' rights, for example, by lengthening periods of variety protection, are ongoing.

When new plant breeds or cultivars are bred, they must be maintained and propagated. Some plants are propagated by asexual means while others are propagated by seeds. Seed propagated cultivars require specific control over seed source and production procedures to maintain the integrity of the plant breeds results. Isolation is necessary to prevent cross contamination with related plants or the mixing of seeds after harvesting. Isolation is normally accomplished by planting distance but in certain crops, plants are enclosed in greenhouses or cages (most commonly used when producing F1 hybrids.)

Role of Plant Breeding in Organic Agriculture

Critics of organic agriculture claim it is too low-yielding to be a viable alternative to conventional agriculture. However, part of that poor performance may be the result of growing poorly adapted varieties. It is estimated that over 95% of organic agriculture is based on conventionally adapted varieties, even though the production environments found in organic vs. conventional farming systems are vastly different due to their distinctive management practices. Most notably, organic farmers have fewer inputs available than conventional growers to control their production environments. Breeding varieties specifically adapted to the unique conditions of organic agriculture is critical for this sector to realize its full potential. This requires selection for traits such as:

- Water use efficiency

- Nutrient use efficiency (particularly nitrogen and phosphorus)

- Weed competitiveness

- Tolerance of mechanical weed control

- Pest/disease resistance

- Early maturity (as a mechanism for avoidance of particular stresses)

- Abiotic stress tolerance (i.e. drought, salinity, etc...)

Currently, few breeding programs are directed at organic agriculture and until recently those that did address this sector have generally relied on indirect selection (i.e. selection in conventional environments for traits considered important for organic agriculture). However, because the difference between organic and conventional environments is large, a given genotype may perform very differently in each environment due to an interaction between genes and the environment. If this interaction is severe enough, an important trait required for the organic environment may not be revealed in the conventional environment, which can result in the selection of poorly adapted individuals. To ensure the most adapted varieties are identified, advocates of organic breeding now promote the use of direct selection (i.e. selection in the target environment) for many agronomic traits.

There are many classical and modern breeding techniques that can be utilized for crop improvement in organic agriculture despite the ban on genetically modified organisms. For instance, controlled crosses between individuals allow desirable genetic variation to be recombined and transferred to seed progeny via natural processes. Marker assisted selection can also be employed as a diagnostics tool to facilitate selection of progeny who possess the desired trait(s), greatly speeding up the breeding process. This technique has proven particularly useful for the introgression of resistance genes into new backgrounds, as well as the efficient selection of many resistance genes pyramided into a single individual. Unfortunately, molecular markers are not currently available for many important traits, especially complex ones controlled by many genes.

Addressing Global Food Security Through Plant Breeding

For future agriculture to thrive there are necessary changes which must be made in accordance to arising global issues. These issues are arable land, harsh cropping conditions and food security which involves, being able to provide the world population with food containing sufficient nutrients. These crops need to be able to mature in several environments allowing for worldwide access, this is involves issues such as drought tolerance. These global issues are achievable through the process of plant breeding, as it offers the ability to select specific genes allowing the crop to perform at a level which yields the desired results.

Minimal Land Degradation

Land degradation is a major issue, as it can negatively impact the capability of the land to be productive. Poor agricultural management has a huge impact on the degradation of soil worldwide and it is Africa and Asia that are most affected. Through education and development of modified plants, these statistics can be reduced and agricultural land can become more productive. Plant breeding allows for an increase in yield with out the extra strain on the land. The genetically modified, Bt white maize, was introduced to South Africa and was surveyed in 33 large commercial farms and 368 small landholders properties and in both cases a higher yield was recorded.

Increased Yield Without Expansion

With an increasing population, the production of food needs to increase with it. It is estimated that a 70% increase in food production is needed by 2050 in order to meet the Declaration of the World Summit on Food Security. But with the natural degradation of agricultural land, simply planting more crops is no longer a viable option. Therefore, new varieties of plants need to be developed through plant breeding that generates an increase of yield without relying on an increase in land area. An example of this can be seen in Asia, where food production per capita has increased twofold. This has been achieved through not only the use of fertilisers, but through the use of better crops that have been specifically designed for the area.

Breeding for Increased Nutritional Value

Plant breeding can contribute to global food security as it is a cost-effective tool for increasing nutritional value of forage and crops. Improvements in nutritional value for forage crops from the use of analytical chemistry and rumen fermentation technology have been recorded since 1960; this science and technology gave breeders the ability to screen thousands of samples within a small amount of time, meaning breeders could identify a high performing hybrid quicker. The main area genetic increases were made was in vitro dry matter digestibility (IVDMD) resulting in 0.7-2.5% increase, at just 1% increase in IVDMD a single Bos Taurus also known as beef cattle reported 3.2% increase in daily gains. This improvement indicates plant breeding is an essential tool in gearing future agriculture to perform at a more advanced level.

Breeding for Tolerance

Plant breeding of hybrid crops has become extremely popular worldwide in an effort to combat the harsh environment. With long periods of drought and lack of water or nitrogen stress tolerance has become a significant part of agriculture. Plant breeders have focused on identifying crops which will ensure crops perform under these conditions; a way to achieve this is finding strains of the crop that is resistance to drought conditions with low nitrogen. It is evident from this that plant breeding is vital for future agriculture to survive as it enables farmers to produce stress resistant crops hence improving food security.

Participatory Plant Breeding

The development of agricultural science, with phenomenon like the Green Revolution arising, have left millions of farmers in developing countries, most of whom operate small farms under unstable and difficult growing conditions, in a precarious situation. The adoption of new plant varieties by this group has been hampered by the constraints of poverty and the international policies promoting an industrialized model of agriculture. Their response has been the creation of a novel and promising set of research methods collectively known as participatory plant breeding. Participatory means that farmers are more involved in the breeding process and breeding goals are defined by farmers instead of international seed companies with their large-scale breeding programs. Farmers' groups and NGOs, for example, may wish to affirm local people's rights over genetic resources, produce seeds themselves, build farmers' technical expertise, or develop new products for niche markets, like organically grown food.

List of Notable Plant Breeders

- Gartons Agricultural Plant Breeders
- Gregor Mendel
- Keith Downey
- Luther Burbank
- Nazareno Strampelli
- Niels Ebbesen Hansen
- Edger McFadden
- Norman Borlaug

Sowing

Simon Bening, *Labors of the Months: September*, from a Flemish Book of hours (Bruges)

Men sowing seed by hand in the 1940s

Sowing is the process of planting seeds. An area or object that has had seeds planted will be described as being sowed.

Traditional Sowing Methods

Plants Which are Usually Sown

Among the major field crops, oats, wheat, and rye are sown, grasses and legumes are seeded, and maize and soybeans are planted. In planting, wider rows (generally 75 cm (30 in) or more) are used, and the intent is to have precise, even spacing between individual seeds in the row; various mechanisms have been devised to count out individual seeds at exact intervals.

Sowing Depth

In sowing, little if any soil is placed over the seeds. More precisely, seeds can be generally sown into the soil by maintaining a planting depth of about 2-3 times the size of the seed.

Regular rows of maize in a field in Indiana.

Sowing Types and Patterns

For hand sowing, several sowing types exist; these include:

- Flat sowing

- Ridge sowing

- Wide bed sowing

Several patterns for sowing may be used together with these types; these include:

- Regular rows

- Rows that are indented at the even rows (so that the seeds are placed in a crossed pattern). This method is much better, as more light may fall on the seedlings as they come out.

- Symmetrical grid pattern – using the quincunx pattern described in The Garden of Cyrus.

Types of Sowing

Hand sowing

Hand sowing or (planting) is the process of casting handfuls of seed over prepared ground, or broadcasting (from which the technological term is derived). Usually, a drag or harrow is employed to incorporate the seed into the soil. Though labor-intensive for any but small areas, this method is still used in some situations. Practice is required to sow evenly and at the desired rate. A hand seeder can be used for sowing, though it is less of a help than it is for the smaller seeds of grasses and legumes.

A tray used in horticulture (for sowing and taking plant cuttings)

Hand sowing may be combined with pre-sowing in seed trays. This allows the plants to come to strength indoors during cold periods (e.g. spring in temperate countries).

In agriculture, most seed is now sown using a seed drill, which offers greater precision; seed is sown evenly and at the desired rate. The drill also places the seed at a measured distance below the soil, so that less seed is required. The standard design uses a fluted feed metering system, which is volumetric in nature; individual seeds are not counted. Rows are typically about 10–30 cm apart, depending on the crop species and growing conditions. Several row opener types are used depending on soil type and local tradition. Grain drills are most often drawn by tractors, but can also be pulled by horses. Pickup trucks are sometimes used, since little draft is required.

A seed rate of about 100 kg of seed per hectare (2 bushels per acre) is typical, though rates vary considerably depending on crop species, soil conditions, and farmer's preference. Excessive rates can cause the crop to lodge, while too thin a rate will result in poor utilisation of the land, competition with weeds and a reduction in the yield.

Open Field

Open-field planting refers to the form of sowing used historically in the agricultural context whereby fields are prepared generically and left open, as the name suggests, before being sown directly with seed. The seed is frequently left uncovered at the surface of the soil before germinating and therefore exposed to the prevailing climate and conditions like storms etc. This is in contrast to the seedbed method used more commonly in domestic gardening or more specific (modern) agricultural scenarios where the seed is applied beneath the soil surface and monitored and manually tended frequently to ensure more successful growth rates and better yields.

Pre-treatment of Seed and Soil Before Sowing

Before sowing, certain seeds first require a treatment prior to the sowing process. This treatment may be seed scarification, stratification, seed soaking or seed cleaning with cold (or medium hot) water.

Tropical fruit such as avocado also benefit from special seed treatments
(specifically invented for that particular tropical fruit)

Seed soaking is generally done by placing seeds in medium hot water for at least 24 to up to 48 hours Seed cleaning is done especially with fruit, as the flesh of the fruit around the seed can quickly become prone to attack from insects or plagues. To clean the seed, usually seed rubbings with cloth/paper is performed, sometimes assisted with a seed washing. Seed washing is generally done by submerging cleansed seeds 20 minutes in 50 degree Celsius water. This (rather hot than moderately hot) water kills any organisms that may have survived on the skin of a seed. Especially with easily infected tropical fruit such as lychees and rambutans, seed washing with high temperature water is vital.

In addition to the mentioned seed pretreatments, seed germination is also assisted when disease-free soil is used. Especially when trying to germinate difficult seed (e.g. certain tropical fruit), prior treatment of the soil (along with the usage of the most suitable soil; e.g. potting soil, prepared soil or other substrates) is vital. The two most used soil treatments are pasteurisation and sterilisation. Depending on the necessity, pasteurisation is to be preferred as this does not kill all organisms. Sterilisation can be done when trying to grow truly difficult crops. To pasteurise the soil, the soil is heated for 15 minutes in an oven of 120 °C.

Weed Control

Weed control is the botanical component of pest control, which attempts to stop weeds, especially noxious or injurious weeds, from competing with domesticated plants and livestock. Many strategies have been developed in order to contain these plants.

The original strategy was manual removal including ploughing, which can cut the roots of weeds. More recent approaches include herbicides (chemical weed killers) and reducing stocks by burning and/or pulverizing seeds.

A plant is often termed a "weed" when it has one or more of the following characteristics:

- Little or no recognized value (as in medicinal, material, nutritional or energy)

- Rapid growth and/or ease of germination

- Competitive with crops for space, light, water and nutrients

The definition of a weed is completely context-dependent. To one person, one plant may be a weed, and to another person it may be a desirable plant. In one place, a plant may be viewed as a weed, whereas in another place, the same plant may be desirable.

Introduction

Weeds compete with productive crops or pasture, ultimately converting productive land into unusable scrub. Weeds can be poisonous, distasteful, produce burrs, thorns or otherwise interfere with the use and management of desirable plants by contaminating harvests or interfering with livestock.

Weeds compete with crops for space, nutrients, water and light. Smaller, slower growing seedlings are more susceptible than those that are larger and more vigorous. Onions are one of the most vulnerable, because they are slow to germinate and produce slender, upright stems. By contrast broad beans produce large seedlings and suffer far fewer effects other than during periods of water shortage at the crucial time when the pods are filling out. Transplanted crops raised in sterile soil or potting compost gain a head start over germinating weeds.

Weeds also vary in their competitive abilities and according to conditions and season. Tall-growing vigorous weeds such as fat hen (*Chenopodium album*) can have the most pronounced effects on adjacent crops, although seedlings of fat hen that appear in late summer produce only small plants. Chickweed (*Stellaria media*), a low growing plant, can happily co-exist with a tall crop during the summer, but plants that have overwintered will grow rapidly in early spring and may swamp crops such as onions or spring greens.

The presence of weeds does not necessarily mean that they are damaging a crop, especially during the early growth stages when both weeds and crops can grow without interference. However, as growth proceeds they each begin to require greater amounts of water and nutrients. Estimates suggest that weed and crop can co-exist harmoniously for around three weeks before competition becomes significant. One study found that after competition had started, the final yield of onion bulbs was reduced at almost 4% per day.

Perennial weeds with bulbils, such as lesser celandine and oxalis, or with persistent underground stems such as couch grass (*Agropyron repens*) or creeping buttercup (*Ranunculus repens*) store reserves of food, and are thus able to grow faster and with more vigour than their annual counterparts. Some perennials such as couch grass exude allelopathic chemicals that inhibit the growth of other nearby plants.

Weeds can also host pests and diseases that can spread to cultivated crops. Charlock and Shepherd's purse may carry clubroot, eelworm can be harboured by chickweed, fat hen and shepherd's purse, while the cucumber mosaic virus, which can devastate the cucurbit family, is carried by a range of different weeds including chickweed and groundsel.

Insect pests often do not attack weeds. However pests such as cutworms may first attack weeds then move on to cultivated crops.

Some plants are considered weeds by some farmers and crops by others. Charlock, a common weed in the southeastern US, are weeds according to row crop growers, but are valued by beekeepers, who seek out places where it blooms all winter, thus providing pollen for honeybees and other pollinators. Its bloom resists all but a very hard freeze, and recovers once the freeze ends.

Weed Propagation

Seeds

Annual and biennial weeds such as chickweed, annual meadow grass, shepherd's purse, groundsel, fat hen, cleaver, speedwell and hairy bittercress propagate themselves by seeding. Many produce huge numbers of seed several times a season, some all year round. Groundsel can produce 1000 seed, and can continue right through a mild winter, whilst Scentless Mayweed produces over 30,000 seeds per plant. Not all of these will germinate at once, but over several seasons, lying dormant in the soil sometimes for years until exposed to light. Poppy seed can survive 80–100 years, dock 50 or more. There can be many thousands of seeds in a square foot or square metre of ground, thus and soil disturbance will produce a flush of fresh weed seedlings.

Subsurface/Surface

The most persistent perennials spread by underground creeping rhizomes that can regrow from a tiny fragment. These include couch grass, bindweed, ground elder, nettles, rosebay willow herb, Japanese knotweed, horsetail and bracken, as well as creeping thistle, whose tap roots can put out lateral roots. Other perennials put out runners that spread along the soil surface. As they creep they set down roots, enabling them to colonise bare ground with great rapidity. These include creeping buttercup and ground ivy. Yet another group of perennials propagate by stolons- stems that arch back into the ground to reroot. The most familiar of these is the bramble.

Methods

Pesticide-free thermic weed control with a weed burner on a potato field in Dithmarschen

Weed control plans typically consist of many methods which are divided into biological, chemical, cultural, and physical/mechanical control.

Physical/Mechanical Methods

Coverings

In domestic gardens, methods of weed control include covering an area of ground with a material that creates a hostile environment for weed growth, known as a *weed mat.*

Several layers of wet newspaper prevent light from reaching plants beneath, which kills them. Daily saturating the newspaper with water plant decomposition. After several weeks, all germinating weed seeds are dead.

In the case of black plastic, the greenhouse effect kills the plants. Although the black plastic sheet is effective at preventing weeds that it covers, it is difficult to achieve complete coverage. Eradicating persistent perennials may require the sheets to be left in place for at least two seasons.

Some plants are said to produce root exudates that suppress herbaceous weeds. *Tagetes minuta* is claimed to be effective against couch and ground elder, whilst a border of comfrey is also said to act as a barrier against the invasion of some weeds including couch. A 5–10 centimetres (2.0–3.9 in)} layer of wood chip mulch prevents most weeds from sprouting.

Gravel can serve as an inorganic mulch.

Irrigation is sometimes used as a weed control measure such as in the case of paddy fields to kill any plant other than the water-tolerant rice crop.

Manual Removal

Weeds are removed manually in large parts of India.

Many gardeners still remove weeds by manually pulling them out of the ground, making sure to include the roots that would otherwise allow them to resprout.

Hoeing off weed leaves and stems as soon as they appear can eventually weaken and kill perennials, although this will require persistence in the case of plants such as bindweed. Nettle infestations can be tackled by cutting back at least three times a year, repeated over a three-year period. Bramble can be dealt with in a similar way.

Tillage

Ploughing includes tilling of soil, intercultural ploughing and summer ploughing. Ploughing uproots weeds, causing them to die. In summer ploughing is done during deep summers. Summer ploughing also helps in killing pests.

Mechanical tilling can remove weeds around crop plants at various points in the growing process.

Thermal

Several thermal methods can control weeds.

Hot foam (foamstream) causes the cell walls to rupture, killing the plant. Weed burners heat up soil quickly and destroy superficial parts of the plants. Weed seeds are often heat resistant and even react with an increase of growth on dry heat.

Since the 19th century soil steam sterilization has been used to clean weeds completely from soil. Several research results confirm the high effectiveness of humid heat against weeds and its seeds.

Soil solarization in some circumstances is very effective at eliminating weeds while maintaining grass. Planted grass tends to have a higher heat/humidity tolerance than unwanted weeds.

Boiling water applied directly to the crown of weeds can also be an effective small weed killer. Larger weeds require three to four applications before being effective.

Seed Targeting

In 1998, the Australian Herbicide Resistance Initiative (AHRI), debuted. gathered fifteen scientists and technical staff members to conduct field surveys, collect seeds, test for resistance and study the biochemical and genetic mechanisms of resistance. A collaboration with DuPont led to a mandatory herbicide labeling program, in which each mode of action is clearly identified by a letter of the alphabet.

The key innovation of the AHRI approach has been to focus on weed seeds. Ryegrass seeds last only a few years in soil, so if farmers can prevent new seeds from arriving, the number of sprouts will shrink each year. Until the new approach farmers were unintentionally helping the seeds. Their combines loosen ryegrass seeds from their stalks and spread them over the fields. In the mid-1980s, a few farmers hitched covered trailers, called "chaff carts", behind their combines to catch the chaff and weed seeds. The collected material is then burned.

An alternative is to concentrate the seeds into a half-meter-wide strip called a windrow and burn the windrows after the harvest, destroying the seeds. Since 2003, windrow burning has been adopted by about 70% of farmers in Western Australia.

Yet another approach is the Harrington Seed Destructor, which is an adaptation of a coal pulverizing cage mill that uses steel bars whirling at up to 1500 rpm. It keeps all the organic material in the field and does not involve combustion, but kills 95% of seeds.

Cultural Methods

Stale Seed Bed

Another manual technique is the 'stale seed bed', which involves cultivating the soil, then leaving it fallow for a week or so. When the initial weeds sprout, the grower lightly hoes them away before planting the desired crop. However, even a freshly cleared bed is susceptible to airborne seed from elsewhere, as well as seed carried by passing animals on their fur, or from imported manure.

Buried Drip Irrigation

Buried drip irrigation involves burying drip tape in the subsurface near the planting bed, thereby limiting weeds access to water while also allowing crops to obtain moisture. It is most effective during dry periods.

Crop Rotation

Rotating crops with ones that kill weeds by choking them out, such as hemp, *Mucuna pruriens*, and other crops, can be a very effective method of weed control. It is a way to avoid the use of herbicides, and to gain the benefits of crop rotation.

Biological Methods

A biological weed control regiment can consist of biological control agents, bioherbicides, use of grazing animals, and protection of natural predators.

Animal Grazing

Companies using goats to control and eradicate leafy spurge, knapweed, and other toxic weeds have sprouted across the American West.

Chemical Methods

"Organic" Approaches

Organic weed control involves anything other than applying manufactured chemicals. Typically a combination of methods are used to achieve satisfactory control.

Weed control, circa 1930-40s

A mechanical weed control device: the diagonal weeder

Sulfur in some circumstances is accepted within British Soil Association standards.

Herbicides

The above described methods of weed control use no or very limited chemical inputs. They are preferred by organic gardeners or organic farmers.

However weed control can also be achieved by the use of herbicides. Selective herbicides kill certain targets while leaving the desired crop relatively unharmed. Some of these act by interfering with the growth of the weed and are often based on plant hormones. Herbicides are generally classified as follows:

- Contact herbicides destroy only plant tissue that contacts the herbicide. Generally, these are the fastest-acting herbicides. They are ineffective on perennial plants that can re-grow from roots or tubers.

- Systemic herbicides are foliar-applied and move through the plant where they destroy a greater amount of tissue. Glyphosate is currently the most used systemic herbicide.

- Soil-borne herbicides are applied to the soil and are taken up by the roots of the target plant.

- Pre-emergent herbicides are applied to the soil and prevent germination or early growth of weed seeds.

In agriculture large scale and systematic procedures are usually required, often by machines, such as large liquid herbicide 'floater' sprayers, or aerial application.

Bradley Method

Bradley Method of Bush Regeneration, which uses ecological processes to do much of the work. Perennial weeds also propagate by seeding; the airborne seed of the dandelion and the rose-bay willow herb parachute far and wide. Dandelion and dock also put down deep tap roots, which, although they do not spread underground, are able to regrow from any remaining piece left in the ground.

Hybrid

One method of maintaining the effectiveness of individual strategies is to combine them with others that work in complete different ways. Thus seed targeting has been combined with herbicides. In Australia seed management has been effectively combined with trifluralin and clethodim.

Resistance

Resistance occurs when a target adapts to circumvent a particular control strategy. It affects not only weed control, but antibiotics, insect control and other domains. In agriculture is mostly considered in reference to pesticides, but can defeat other strategies, e.g., when a target species becomes more drought tolerant via selection pressure.

Farming practices

Herbicide resistance recently became a critical problem as many Australian sheep farmers switched to exclusively growing wheat in their pastures in the 1970s. In wheat fields, introduced varieties of ryegrass, while good for grazing sheep, are intense competitors with wheat. Ryegrasses produce so many seeds that, if left unchecked, they can completely choke a field. Herbicides provided excellent control, while reducing soil disrupting because of less need to plough. Within little more than a decade, ryegrass and other weeds began to develop resistance. Australian farmers evolved again and began diversifying their techniques.

In 1983, patches of ryegrass had become immune to Hoegrass, a family of herbicides that inhibit an enzyme called acetyl coenzyme A carboxylase.

Ryegrass populations were large, and had substantial genetic diversity, because farmers had planted many varieties. Ryegrass is cross-pollinated by wind, so genes shuffle frequently. Farmers sprayed inexpensive Hoegrass year after year, creating selection pressure, but were diluting the herbicide in order to save money, increasing plants survival. Hoegrass was mostly replaced by a group of herbicides that block acetolactate synthase, again helped by poor application practices. Ryegrass evolved a kind of "cross-resistance" that allowed it to rapidly break down a variety of herbicides. Australian farmers lost four classes of herbicides in only a few years. As of 2013 only two herbicide classes, called Photosystem II and long-chain fatty acid inhibitors, had become the last hope.

References

- Bleasdale, J. K. A.; Salter, Peter John (1 January 1991). The Complete Know and Grow Vegetables. Oxford University Press. ISBN 978-0-19-286114-6.

- Ross, Merrill A.; Lembi, Carole A. (2008). Applied Weed Science: Including the Ecology and Management of Invasive Plants. Prentice Hall. p. 123. ISBN 978-0135028148.

- Dinesh, Harshavardhan; Pearce, Joshua M. (2016-02-01). "The potential of agrivoltaic systems". Renewable and Sustainable Energy Reviews 54: 299–308. doi:10.1016/j.rser.2015.10.024.

- Suzie Key; Julian K-C Ma & Pascal MW Drake (1 June 2008). "Genetically modified plants and human health". Journal of the Royal Society of Medecine. pp. 290–298. Retrieved 11 March 2015.

- Teklu, Misghina G.; Schomaker, Corrie H.; Been, Thomas H. (2014-05-28). "Relative susceptibilities of five fodder radish varieties (Raphanus sativus var. Oleiformis) to Meloidogyne chitwoodi". Nematology 16 (5): 577–590. doi:10.1163/15685411-00002789. ISSN 1568-5411.

- Stokstad, E. (2013). "The War Against Weeds Down Under". Science. 341 (6147): 734. doi:10.1126/science.341.6147.734.

- Oldeman, l (1994). "The global extent of soil degradation" (PDF). Soil resilience and sustainable land use. 32 (5967): 818–822. Retrieved 7-11-2013.

- Bänziger (2000). "Breeding for drought and nitrogen stress tolerance in maize: from theory to practice". From Theory to Practice: 7–9. Retrieved 7-11-2013.

- Lithourgidis, A.S.; Dordas, C.A.; Damalas, C.A.; Vlachostergios, D.N. (2011). "Annual intercrops: an alternative pathway for sustainable agriculture" (PDF). Australian Journal of Crop Science 5 (4): 396–410.

4

Integrated Topics of Agricultural Science

Tests and observations required by agricultural science are performed with the help of its integrated branches of study. Some of these fields are agrophysics and agricultural economics. This chapter elucidates the crucial theories and principles of agricultural science.

Agricultural Biotechnology

Agricultural biotechnology, also known as agritech, is an area of agricultural science involving the use of scientific tools and techniques, including genetic engineering, molecular markers, molecular diagnostics, vaccines, and tissue culture, to modify living organisms: plants, animals, and microorganisms. Crop Biotechnology is one aspect of Agricultural Biotechnology which has been greatly developed upon in recent times. Desired trait are exported from a particular species of Crop to an entirely different species. These Transgene crops possess desirable characteristics in terms of flavor, color of flowers, growth rate, size of harvested products and resistance to diseases and pests.

Agrophysics

Agrophysics is a branch of science bordering on agronomy and physics, whose objects of study are the agroecosystem - the biological objects, biotope and biocoenosis affected by human activity, studied and described using the methods of physical sciences. Using the achievements of the exact sciences to solve major problems in agriculture, agrophysics involves the study of materials and processes occurring in the production and processing of agricultural crops, with particular emphasis on the condition of the environment and the quality of farming materials and food production.

Agrophysics is closely related to biophysics, but is restricted to the biology of the plants, animals, soil and an atmosphere involved in agricultural activities and biodiversity. It is different from biophysics in having the necessity of taking into account the specific features of biotope and biocoenosis, which involves the knowledge of nutritional science and agroecology, agricultural technology, biotechnology, genetics etc.

The needs of agriculture, concerning the past experience study of the local complex soil and next plant-atmosphere systems, lay at the root of the emergence of a new branch – agrophysics – dealing this with experimental physics. The scope of the branch starting from soil science (physics) and originally limited to the study of relations within the soil environment, expanded over time onto influencing the properties of agricultural crops and produce as foods and raw postharvest materials, and onto the issues of quality, safety and labeling concerns, considered distinct from the field of nutrition for application in food science.

Research centres focused on the development of the agrophysical sciences include the Institute of Agrophysics, Polish Academy of Sciences in Lublin, and the Agrophysical Research Institute, Russian Academy of Sciences in St. Petersburg.

Research institutes and societies

- Agrophysical Research Institute in St. Petersburg, Russia

- Bohdan Dobrzański Institute of Agrophysics in Lublin, Poland

- The Indian Society of AgroPhysics

Agricultural Chemistry

Agricultural chemistry is the study of both chemistry and biochemistry which are important in agricultural production, the processing of raw products into foods and beverages, and in environmental monitoring and remediation. These studies emphasize the relationships between plants, animals and bacteria and their environment. The science of chemical compositions and changes involved in the production, protection, and use of crops and livestock. As a basic science, it embraces, in addition to test-tube chemistry, all the life processes through which humans obtain food and fiber for themselves and feed for their animals. As an applied science or technology, it is directed toward control of those processes to increase yields, improve quality, and reduce costs. One important branch of it, chemurgy, is concerned chiefly with utilization of agricultural products as chemical raw materials.

Sciences

The goals of agricultural chemistry are to expand understanding of the causes and effects of biochemical reactions related to plant and animal growth, to reveal opportunities for controlling those reactions, and to develop chemical products that will provide the desired assistance or control. Every scientific discipline that contributes to agricultural progress depends in some way on chemistry. Hence agricultural chemistry is not a distinct discipline, but a common thread that ties together genetics, physiology, microbiology, entomology, and numerous other sciences that impinge on agriculture.

Chemical materials developed to assist in the production of food, feed, and fiber include scores of herbicides, insecticides, fungicides, and other pesticides, plant growth regulators, fertilizers, and animal feed supplements. Chief among these groups from the commercial point of view are manufactured fertilizers, synthetic pesticides (including herbicides), and supplements for feeds. The latter include both nutritional supplements (for example, mineral nutrients) and medicinal compounds for the prevention or control of disease.

Agricultural chemistry often aims at preserving or increasing the fertility of soil, maintaining or improving the agricultural yield, and improving the quality of the crop.

When agriculture is considered with ecology, the sustainablility of an operation is considered. Modern agrochemical industry has gained a reputation for maximising profits while violating sustainable and ecologically viable agricultural principles. Eutrophication, the prevalence of genetically modified crops and the increasing concentration of chemicals in the food chain (e.g. persistent organic pollutants) are only a few consequences of naive industrial agriculture.

History

- In 1761 Johan Gottschalk Wallerius publishes his pioneering work, *Agriculturae fundamenta chemica (Åkerbrukets chemiska grunder)*.

- In 1815 Humphry Davy publishes *Elements of agricultural chemistry*

- In 1842 Justus von Liebig publishes *Animal Chemistry or Organic Chemistry in its applications to Physiology and Pathology*.

- Jöns Jacob Berzelius publishes *Traité de chimie minérale, végétale et animal* (6 vols., 1845–50)

- Jean-Baptiste Boussingault publishes *Agronomie, chimie agricole, et physiologie* (5 vols., 1860–1874; 2nd ed., 1884).

- In 1868 Samuel William Johnson publishes *How Crops Grow*.

- In 1870 S. W. Johnson publishes *How Crops Feed: A treatise on the atmosphere and soil as related to the nutrition of agricultural plants*.

- In 1872 Karl Heinrich Ritthausen publishes *Protein bodies in grains, legumes, and linseed. Contributions to the physiology of seeds for cultivation, nutrition, and fodder*

Agricultural Economics

Agricultural economics or agronomics is an applied field of economics concerned with the application of economic theory in optimizing the production and distribution of food and fibre — a discipline known as agronomics. Agronomics was a branch of economics that specifically dealt with land usage. It focused on maximizing the crop yield while maintaining a good soil ecosystem. Throughout the 20th century the discipline expanded and the current scope of the discipline is much broader. Agricultural economics today includes a variety of applied areas, having considerable overlap with conventional economics. Agricultural economists have made substantial contributions to research in economics, econometrics, development economics, and environmental economics. Agricultural economics influences food policy, agricultural policy, and environmental policy.

Origins

Economics has been defined as the study of resource allocation under scarcity. Agronomics, or the application of economic methods to optimizing the decisions made by agricultural producers, grew to prominence around the turn of the 20th century. The field of agricultural economics can be traced out to works on land economics. Henry Charles Taylor was the greatest contributor with the establishment of the Department of Agricultural Economics at Wisconsin in 1909.

Another contributor, 1979 Nobel Economics Prize winner Theodore Schultz, was among the first to examine development economics as a problem related directly to agriculture. Schultz was also instrumental in establishing econometrics as a tool for use in analyzing agricultural economics empirically; he noted in his landmark 1956 article that agricultural supply analysis is rooted in "shifting sand", implying that it was and is simply not being done correctly.

One scholar summarizes the development of agricultural economics as follows:

"Agricultural economics arose in the late 19th century, combined the theory of the firm with marketing and organization theory, and developed throughout the 20th century largely as an empirical branch of general economics. The discipline was closely linked to empirical applications of mathematical statistics and made early and significant contributions to econometric methods. In the 1960s and afterwards, as agricultural sectors in the OECD countries contracted, agricultural economists were drawn to the development problems of poor countries, to the trade and macroeconomic policy implications of agriculture in rich countries, and to a variety of production, consumption, and environmental and resource problems."

Agricultural economists have made many well-known contributions to the economics field with such models as the cobweb model, hedonic regression pricing models, new technology and diffusion models (Zvi Griliches), multifactor productivity and efficiency theory and measurement, and the random coefficients regression. The farm sector is frequently cited as a prime example of the perfect competition economic paradigm.

In Asia, agricultural economics was offered first by the University of the Philippines Los Baños Department of Agricultural Economics in 1919. Today, the field of agricultural economics has transformed into a more integrative discipline which covers farm management and production economics, rural finance and institutions, agricultural marketing and prices, agricultural policy and development, food and nutrition economics, and environmental and natural resource economics.

Since the 1970s, agricultural economics has primarily focused on seven main topics, according to a scholar in the field: agricultural environment and resources; risk and uncertainty; food and consumer economics; prices and incomes; market structures; trade and development; and technical change and human capital.

Major Topics in Agricultural Economics

Agricultural Environment and Natural Resources

In the field of environmental economics, agricultural economists have contributed in three main areas: designing incentives to control environmental externalities (such as water pollution due to agricultural production), estimating the value of non-market benefits from natural resources and environmental amenities (such as an appealing rural landscape), and the complex interrelationship between economic activities and environmental consequences. With regard to natural resources, agricultural economists have developed quantitative tools for improving land management, preventing erosion, managing pests, protecting biodiversity, and preventing livestock diseases.

Food and Consumer Economics

While at one time, the field of agricultural economics was focused primarily on farm-level issues, in recent years agricultural economists have studied diverse topics related to the economics of food consumption. In addition to economists' long-standing emphasis on the effects of prices and incomes, researchers in this field have studied how information and quality attributes influence consumer behavior. Agricultural economists have contributed to understanding how households make choices between purchasing food or preparing it at home, how food prices are determined, definitions of poverty thresholds, how consumers respond to price and income changes in a consistent way, and survey and experimental tools for understanding consumer preferences.

Production Economics and Farm Management

Agricultural economics research has addressed diminishing returns in agricultural production, as well as farmers' costs and supply responses. Much research has applied economic theory to farm-level decisions. Studies of risk and decision-making under uncertainty have real-world applications to crop insurance policies and to understanding how farmers in developing countries make choices about technology adoption. These topics are important for understanding prospects for producing sufficient food for a growing world population, subject to new resource and environmental challenges such as water scarcity and global climate change.

Professional Associations

The International Association of Agricultural Economists (IAAE) is a worldwide professional association, which holds its major conference once every three years. The association publishes the journal *Agricultural Economics*. There also is a European Association of Agricultural Economists (EAAE), African association of agricultural economists[AAAE]and an Australian Agricultural and Resource Economics Society. Substantial work in agricultural economics internationally is conducted by the International Food Policy Research Institute.

In the United States, the primary professional association is the Agricultural & Applied Economics Association (AAEA), which holds its own annual conference and also co-sponsors the annual meetings of the Allied Social Sciences Association (ASSA). The AAEA publishes the American Journal of Agricultural Economics and Applied Economic Perspectives and Policy.

Careers in Agricultural Economics

Graduates from agricultural and applied economics departments find jobs in many sectors of the economy: agricultural management, agribusiness, commodities markets, education, the financial sector, government, natural resource and environmental management, real estate, and public relations. Careers in agricultural economics require at least a bachelor's degree, and research careers in the field require graduate-level training. A 2011 study by the Georgetown Center on Education and the Workforce rated agricultural economics tied for 8th out of 171 fields in terms of employability.

Literature

Evenson, Robert E. and Prabhu Pingali (eds.) (2007). *Handbook of Agricultural Economics*. Amsterdam, NL: Elsevier.

References

- Encyclopedia of Soil Science, edts. Ward Chesworth, 2008, Uniw. of Guelph Canada, Publ. Springer, ISBN 978-1-4020-3994-2

- Scientific Dictionary of Agrophysics: polish-English, polsko-angielski by R. Dębicki, J. Gliński, J. Horabik, R. T. Walczak - Lublin 2004, ISBN 83-87385-88-3

- Physical Methods in Agriculture. Approach to Precision and Quality, edts. J. Blahovec and M. Kutilek, Kluwer Academic Publishers, New York 2002, ISBN 0-306-47430-1.

- Soil Physical Condition and Plant Roots by J. Gliński, J. Lipiec, 1990, CRC Press, Inc., Boca Raton, USA, ISBN 0-8493-6498-1

- Soil Aeration and its Role for Plants by J. Gliński, W. Stępniewski, 1985, Publisher: CRC Press, Inc., Boca Raton, USA, ISBN 0-8493-5250-9

- Department of Agricultural, Environmental, and Development Economics, the Ohio State University (2014). "What's the Value of an Agricultural Economics Degree?". Retrieved 2014-10-11. CS1 maint: Uses authors parameter (link)

- Anthony P. Carnevale; Jeff Strohl; Michelle Melton (2011). "What's It Worth? The Economic Value of College Majors". Retrieved 2014-10-11.

5

Alternative Agricultural Systems

Growing concerns about irreparable environmental damage and pesticide related diseases in humans and animals have highlighted the importance of alternative agricultural systems. Discussed in this chapter are practices like organic farming, rainfed agriculture and cash crops. This chapter discusses the methods of alternative agriculture in a critical manner providing key analysis to the subject matter.

Organic Farming

Vegetables from organic farming.

Organic farming is an alternative agricultural system which originated early in the 20th century in reaction to rapidly changing farming practices. Organic agriculture continues to be developed by various organic agriculture organizations today. It relies on fertilizers of organic origin such as compost, manure, green manure, and bone meal and places emphasis on techniques such as crop rotation and companion planting. Biological pest control, mixed cropping and the fostering of insect predators are encouraged. In general, organic standards are designed to allow the use of naturally occurring substances while prohibiting or strictly limiting synthetic substances. For instance, naturally occurring pesticides such as pyrethrin and rotenone are permitted, while synthetic fertilizers and pesticides are generally prohibited. Synthetic substances that are allowed include, for example, copper sulfate, elemental sulfur and Ivermectin. Genetically modified organisms, nanomaterials, human sewage sludge, plant growth regulators, hormones, and antibiotic use in livestock husbandry are prohibited. Reasons for advocation of organic farming include real or perceived advantages in sustainability, openness, independence, health, food security, and food safety, although the match between perception and reality is continually challenged.

Organic agricultural methods are internationally regulated and legally enforced by many nations, based in large part on the standards set by the International Federation of Organic Agriculture Movements (IFOAM), an international umbrella organization for organic farming organizations established in 1972. Organic agriculture can be defined as:

an integrated farming system that strives for sustainability, the enhancement of soil fertility and biological diversity whilst, with rare exceptions, prohibiting synthetic pesticides, antibiotics, synthetic fertilizers, genetically modified organisms, and growth hormones.

Since 1990 the market for organic food and other products has grown rapidly, reaching $63 billion worldwide in 2012. This demand has driven a similar increase in organically managed farmland that grew from 2001 to 2011 at a compounding rate of 8.9% per annum. As of 2011, approximately 37,000,000 hectares (91,000,000 acres) worldwide were farmed organically, representing approximately 0.9 percent of total world farmland.

History

Agriculture was practiced for thousands of years without the use of artificial chemicals. Artificial fertilizers were first created during the mid-19th century. These early fertilizers were cheap, powerful, and easy to transport in bulk. Similar advances occurred in chemical pesticides in the 1940s, leading to the decade being referred to as the 'pesticide era'. These new agricultural techniques, while beneficial in the short term, had serious longer term side effects such as soil compaction, erosion, and declines in overall soil fertility, along with health concerns about toxic chemicals entering the food supply. In the late 1800s and early 1900s, soil biology scientists began to seek ways to remedy these side effects while still maintaining higher production.

Biodynamic agriculture was the first modern system of agriculture to focus exclusively on organic methods. Its development began in 1924 with a series of eight lectures on agriculture given by Rudolf Steiner. These lectures, the first known presentation of what later came to be known as organic agriculture, were held in response to a request by farmers who noticed degraded soil conditions and a deterioration in the health and quality of crops and livestock resulting from the use of chemical fertilizers. The one hundred eleven attendees, less than half of whom were farmers, came from six countries, primarily Germany and Poland. The lectures were published in November 1924; the first English translation appeared in 1928 as *The Agriculture Course*.

In 1921, Albert Howard and his wife Gabrielle Howard, accomplished botanists, founded an Institute of Plant Industry to improve traditional farming methods in India. Among other things, they brought improved implements and improved animal husbandry methods from their scientific training; then by incorporating aspects of the local traditional methods, developed protocalls for the rotation of crops, erosion prevention techniques, and the systematic use of composts and manures. Stimulated by these experiences of traditional farming, when Albert Howard returned to Britain in the early 1930s he began to promulgate a system of natural agriculture.

In July 1939, Ehrenfried Pfeiffer, the author of the standard work on biodynamic agriculture (*Bio-Dynamic Farming and Gardening*), came to the UK at the invitation of Walter James, 4th Baron Northbourne as a presenter at the Betteshanger Summer School and Conference on Biodynamic Farming at Northbourne's farm in Kent. One of the chief purposes of the conference was to bring together the proponents of various approaches to organic agriculture in order that they might cooperate within a larger movement. Howard attended the conference, where he met Pfeiffer. In the following year, Northbourne published his manifesto of organic farming, *Look to the Land*, in which he coined the term "organic farming." The Betteshanger conference has been described as the 'missing link' between biodynamic agriculture and other forms of organic farming.

In 1940 Howard published his *An Agricultural Testament*. In this book he adopted Northbourne's terminology of "organic farming." Howard's work spread widely, and he became known as the "father of organic farming" for his work in applying scientific knowledge and principles to various traditional and natural methods. In the United States J.I. Rodale, who was keenly interested both in Howard's ideas and in biodynamics, founded in the 1940s both a working organic farm for trials and experimentation, The Rodale Institute, and the Rodale Press to teach and advocate organic methods to the wider public. These became important influences on the spread of organic agriculture. Further work was done by Lady Eve Balfour in the United Kingdom, and many others across the world.

Increasing environmental awareness in the general population in modern times has transformed the originally supply-driven organic movement to a demand-driven one. Premium prices and some government subsidies attracted farmers. In the developing world, many producers farm according to traditional methods that are comparable to organic farming, but not certified, and that may not include the latest scientific advancements in organic agriculture. In other cases, farmers in the developing world have converted to modern organic methods for economic reasons.

Terminology

Biodynamic agriculturists, who based their work on Steiner's spiritually-oriented anthroposophy, used the term "organic" to indicate that a farm should be viewed as a living organism, in the sense of the following quotation:

"An organic farm, properly speaking, is not one that uses certain methods and substances and avoids others; it is a farm whose structure is formed in imitation of the structure of a natural system that has the integrity, the independence and the benign dependence of an organism"

— Wendell Berry, "The Gift of Good Land"

The use of "organic" popularized by Howard and Rodale, on the other hand, refers more narrowly to the use of organic matter derived from plant compost and animal manures to improve the humus content of soils, grounded in the work of early soil scientists who developed what was then called "humus farming." Since the early 1940s the two camps have tended to merge.

Methods

"Organic agriculture is a production system that sustains the health of soils, ecosystems and people. It relies on ecological processes, biodiversity and cycles adapted to local conditions, rather than the use of inputs with adverse effects. Organic agriculture combines tradition, innovation and science to benefit the shared environment and promote fair relationships and a good quality of life for all involved..."

— International Federation of Organic Agriculture Movements

Organic farming methods combine scientific knowledge of ecology and modern technology with traditional farming practices based on naturally occurring biological processes. Organic farming methods are studied in the field of agroecology. While conventional agriculture uses synthetic pesticides and water-soluble synthetically purified fertilizers, organic farmers are restricted by regulations to using natural pesticides and fertilizers. An example of a natural pesticide is pyrethrin, which is found naturally in the Chrysanthemum flower. The principal methods of organic farming include crop rotation, green manures and compost, biological pest control, and mechanical cultivation. These measures use the natural environment to enhance agricultural productivity: legumes are planted to fix nitrogen into the soil, natural insect predators are encouraged, crops are rotated to confuse pests and renew soil, and natural materials such as potassium bicarbonate and mulches are used to control disease and weeds. Genetically modified seeds and animals are excluded.

Organic cultivation of mixed vegetables in Capay, California. Note the hedgerow in the background.

While organic is fundamentally different from conventional because of the use of carbon based fertilizers compared with highly soluble synthetic based fertilizers and biological pest control instead of synthetic pesticides, organic farming and large-scale conventional farming are not entirely mutually exclusive. Many of the methods developed for organic agriculture have been borrowed by more conventional agriculture. For example, Integrated Pest Management is a multifaceted strategy that uses various organic methods of pest control whenever possible, but in conventional farming could include synthetic pesticides only as a last resort.

Crop Diversity

Organic farming encourages Crop diversity. The science of agroecology has revealed the benefits of polyculture (multiple crops in the same space), which is often employed in organic farming. Planting a variety of vegetable crops supports a wider range of beneficial insects, soil microorganisms, and other factors that add up to overall farm health. Crop diversity helps environments thrive and protects species from going extinct.

Soil Management

Organic farming relies heavily on the natural breakdown of organic matter, using techniques like green manure and composting, to replace nutrients taken from the soil by previous crops. This biological process, driven by microorganisms such as mycorrhiza, allows the natural production of nutrients in the soil throughout the growing season, and has been referred to as *feeding the soil to feed the plant*. Organic farming uses a variety of methods to improve soil fertility, including crop rotation, cover cropping, reduced tillage, and application of compost. By reducing tillage, soil is not inverted and exposed to air; less carbon is lost to the atmosphere resulting in more soil organic carbon. This has an added benefit of carbon sequestration, which can reduce green house gases and help reverse climate change.

Plants need nitrogen, phosphorus, and potassium, as well as micronutrients and symbiotic relationships with fungi and other organisms to flourish, but getting enough nitrogen, and particularly synchronization so that plants get enough nitrogen at the right time (when plants need it most), is a challenge for organic farmers. Crop rotation and green manure ("cover crops") help to provide nitrogen through legumes (more precisely, the *Fabaceae* family), which fix nitrogen from the atmosphere through symbiosis with rhizobial bacteria. Intercropping, which is sometimes used for insect and disease control, can also increase soil nutrients, but the competition between the legume and the crop can be problematic and wider spacing between crop rows is required. Crop residues can be ploughed back into the soil, and different plants leave different amounts of nitrogen, potentially aiding synchronization. Organic farmers also use animal manure, certain processed fertilizers such as seed meal and various mineral powders such as rock phosphate and green sand, a naturally occurring form of potash that provides potassium. Together these methods help to control erosion. In some cases pH may need to be amended. Natural pH amendments include lime and sulfur, but in the U.S. some compounds such as iron sulfate, aluminum sulfate, magnesium sulfate, and soluble boron products are allowed in organic farming.

Mixed farms with both livestock and crops can operate as ley farms, whereby the land gathers fertility through growing nitrogen-fixing forage grasses such as white clover or alfalfa and grows cash crops or cereals when fertility is established. Farms without livestock ("stockless") may find it more difficult to maintain soil fertility, and may rely more on external inputs such as imported manure as well as grain legumes and green manures, although grain legumes may fix limited nitrogen because they are harvested. Horticultural farms that grow fruits and vegetables in protected conditions often relay even more on external inputs.

Biological research into soil and soil organisms has proven beneficial to organic farming. Varieties of bacteria and fungi break down chemicals, plant matter and animal waste into productive soil nutrients. In turn, they produce benefits of healthier yields and more productive soil for future crops. Fields with less or no manure display significantly lower yields, due to decreased soil microbe community. Increased manure improves biological activity, providing a healthier, more arable soil system and higher yields.

Weed Management

Organic weed management promotes weed suppression, rather than weed elimination, by enhancing crop competition and phytotoxic effects on weeds. Organic farmers integrate cultural, biological, mechanical, physical and chemical tactics to manage weeds without synthetic herbicides.

Organic standards require rotation of annual crops, meaning that a single crop cannot be grown in the same location without a different, intervening crop. Organic crop rotations frequently include weed-suppressive cover crops and crops with dissimilar life cycles to discourage weeds associated with a particular crop. Research is ongoing to develop organic methods to promote the growth of natural microorganisms that suppress the growth or germination of common weeds.

Other cultural practices used to enhance crop competitiveness and reduce weed pressure include selection of competitive crop varieties, high-density planting, tight row spacing, and late planting into warm soil to encourage rapid crop germination.

Mechanical and physical weed control practices used on organic farms can be broadly grouped as:

- Tillage - Turning the soil between crops to incorporate crop residues and soil amendments; remove existing weed growth and prepare a seedbed for planting; turning soil after seeding to kill weeds, including cultivation of row crops;

- Mowing and cutting - Removing top growth of weeds;

- Flame weeding and thermal weeding - Using heat to kill weeds; and

- Mulching - Blocking weed emergence with organic materials, plastic films, or landscape fabric.

Some critics, citing work published in 1997 by David Pimentel of Cornell University, which described an epidemic of soil erosion worldwide, have raised concerned that tillage contribute to the erosion epidemic. The FAO and other organizations have advocated a 'no-till' approach to both conventional and organic farming, and point out in particular that crop rotation techniques used in organic farming are excellent no-till approaches. A study published in 2005 by Pimentel and colleagues confirmed that 'Crop rotations and cover cropping (green manure) typical of organic agriculture reduce soil erosion, pest problems, and pesticide use.' Some naturally sourced chemicals are allowed for herbicidal use. These include certain formulations of acetic acid (concentrated vinegar), corn gluten meal, and essential oils. A few selective bioherbicides based on fungal pathogens have also been developed. At this time, however, organic herbicides and bioherbicides play a minor role in the organic weed control toolbox.

Weeds can be controlled by grazing. For example, geese have been used successfully to weed a range of organic crops including cotton, strawberries, tobacco, and corn, reviving the practice of keeping cotton patch geese, common in the southern U.S. before the 1950s. Similarly, some rice farmers introduce ducks and fish to wet paddy fields to eat both weeds and insects.

Controlling other organisms

Organisms aside from weeds that cause problems on organic farms include arthropods (e.g., insects, mites), nematodes, fungi and bacteria. Organic practices include, but are not limited to:

- encouraging predatory beneficial insects to control pests by serving them nursery plants and/or an alternative habitat, usually in a form of a shelterbelt, hedgerow, or beetle bank;

- encouraging beneficial microorganisms;

- rotating crops to different locations from year to year to interrupt pest reproduction cycles;

- planting companion crops and pest-repelling plants that discourage or divert pests;

- using row covers to protect crops during pest migration periods;

- using biologic pesticides and herbicides

- using stale seed beds to germinate and destroy weeds before planting

- using sanitation to remove pest habitat;

- Using insect traps to monitor and control insect populations.

- Using physical barriers, such as row covers

Chloroxylon is used for Pest Management in Organic Rice Cultivation in Chhattisgarh, India

Examples of predatory beneficial insects include minute pirate bugs, big-eyed bugs, and to a lesser extent ladybugs (which tend to fly away), all of which eat a wide range of pests. Lacewings are also effective, but tend to fly away. Praying mantis tend to move more slowly and eat less heavily. Parasitoid wasps tend to be effective for their selected prey, but like all small insects can be less effective outdoors because the wind controls their movement. Predatory mites are effective for controlling other mites.

Naturally derived insecticides allowed for use on organic farms use include *Bacillus thuringiensis* (a bacterial toxin), pyrethrum (a chrysanthemum extract), spinosad (a bacterial metabolite), neem (a tree extract) and rotenone (a legume root extract). Fewer than 10% of organic farmers use these pesticides regularly; one survey found that only 5.3% of vegetable growers in California use rotenone while 1.7% use pyrethrum. These pesticides are not always more safe or environmentally friendly than synthetic pesticides and can cause harm. The main criterion for organic pesticides is that they are naturally derived, and some naturally derived substances have been controversial. Controversial natural pesticides include rotenone, copper, nicotine sulfate, and pyrethrums Rotenone and pyrethrum are particularly controversial because they work by attacking the nervous system, like most conventional insecticides. Rotenone is extremely toxic to fish and can induce symptoms resembling Parkinson's disease in mammals. Although pyrethrum (natural pyrethrins) is more effective against insects when used with piperonyl butoxide (which retards degradation of the pyrethrins), organic standards generally do not permit use of the latter substance.

Naturally derived fungicides allowed for use on organic farms include the bacteria *Bacillus subtilis* and *Bacillus pumilus*; and the fungus *Trichoderma harzianum*. These are mainly effective for diseases affecting roots. Compost tea contains a mix of beneficial microbes, which may attack or out-compete certain plant pathogens, but variability among formulations and preparation methods may contribute to inconsistent results or even dangerous growth of toxic microbes in compost teas.

Some naturally derived pesticides are not allowed for use on organic farms. These include nicotine sulfate, arsenic, and strychnine.

Synthetic pesticides allowed for use on organic farms include insecticidal soaps and horticultural oils for insect management; and Bordeaux mixture, copper hydroxide and sodium bicarbonate for managing fungi. Copper sulfate and Bordeaux mixture (copper sulfate plus lime), approved for organic use in various jurisdictions, can be more environmentally problematic than some synthetic fungicides dissallowed in organic farming Similar concerns apply to copper hydroxide. Repeated application of copper sulfate or copper hydroxide as a fungicide may eventually result in copper accumulation to toxic levels in soil, and admonitions to avoid excessive accumulations of copper in soil appear in various organic standards and elsewhere. Environmental concerns for several kinds of biota arise at average rates of use of such substances for some crops. In the European Union, where replacement of copper-based fungicides in organic agriculture is a policy priority, research is seeking alternatives for organic production.

Livestock

For livestock like these healthy cows vaccines play an important part in animal health since antibiotic therapy is prohibited in organic farming

Raising livestock and poultry, for meat, dairy and eggs, is another traditional farming activity that complements growing. Organic farms attempt to provide animals with natural living conditions and feed. Organic certification verifies that livestock are raised according to the USDA organic regulations throughout their lives. These regulations include the requirement that all animal feed must be certified organic.

Organic livestock may be, and must be, treated with medicine when they are sick, but drugs cannot be used to promote growth, their feed must be organic, and they must be pastured.

Also, horses and cattle were once a basic farm feature that provided labor, for hauling and plowing, fertility, through recycling of manure, and fuel, in the form of food for farmers and other animals. While today, small growing operations often do not include livestock, domesticated animals are a desirable part of the organic farming equation, especially for true sustainability, the ability of a farm to function as a self-renewing unit.

Genetic Modification

A key characteristic of organic farming is the rejection of genetically engineered plants and animals. On 19 October 1998, participants at IFOAM's 12th Scientific Conference issued the Mar del Plata Declaration, where more than 600 delegates from over 60 countries voted unanimously to exclude the use of genetically modified organisms in food production and agriculture.

Although opposition to the use of any transgenic technologies in organic farming is strong, agricultural researchers Luis Herrera-Estrella and Ariel Alvarez-Morales continue to advocate integration of transgenic technologies into organic farming as the optimal means to sustainable agriculture, particularly in the developing world, as does author and scientist Pamela Ronald, who views this kind of biotechnology as being consistent with organic principles.

Although GMOs are excluded from organic farming, there is concern that the pollen from genetically modified crops is increasingly penetrating organic and heirloom seed stocks, making it difficult, if not impossible, to keep these genomes from entering the organic food supply. Differing regulations among countries limits the availability of GMOs to certain countries, as described in the article on regulation of the release of genetic modified organisms.

Tools

Organic farmers use a number of traditional farm tools to do farming. Due to the goals of sustainability in organic farming, organic farmers try to minimize their reliance on fossil fuels. In the developing world on small organic farms tools are normally constrained to hand tools and diesel powered water pumps. Some organic farmers make use of renewable energy on the farm and can even make use of agrivoltaics or other onsite colocation of power production and agriculture. A recent study evaluated the use of open-source 3-D printers (called RepRaps using a bioplastic polylactic acid (PLA) on organic farms. PLA is a strong biodegradable and recyclable thermoplastic appropriate for a range of representative products in five categories of prints: handtools, food processing, animal management, water management and hydroponics. Such open source hardware is attractive to all types of small farmers as it provides control for farmers over their own equipment; this is exemplified by Open Source Ecology, Farm Hack and Farmbot.io.

Standards

Standards regulate production methods and in some cases final output for organic agriculture.

Standards may be voluntary or legislated. As early as the 1970s private associations certified organic producers. In the 1980s, governments began to produce organic production guidelines. In the 1990s, a trend toward legislated standards began, most notably with the 1991 EU-Eco-regulation developed for European Union, which set standards for 12 countries, and a 1993 UK program. The EU's program was followed by a Japanese program in 2001, and in 2002 the U.S. created the National Organic Program (NOP). As of 2007 over 60 countries regulate organic farming (IFOAM 2007). In 2005 IFOAM created the Principles of Organic Agriculture, an international guideline for certification criteria. Typically the agencies accredit certification groups rather than individual farms.

Organic production materials used in and foods are tested independently by the Organic Materials Review Institute.

Composting

Using manure as a fertiliser risks contaminating food with animal gut bacteria, including pathogenic strains of E. coli that have caused fatal poisoning from eating organic food. To combat this risk, USDA organic standards require that manure must be sterilized through high temperature thermophilic composting. If raw animal manure is used, 120 days must pass before the crop is harvested if the final product comes into direct contact with the soil. For products that don't directly contact soil, 90 days must pass prior to harvest.

Economics

The economics of organic farming, a subfield of agricultural economics, encompasses the entire process and effects of organic farming in terms of human society, including social costs, opportunity costs, unintended consequences, information asymmetries, and economies of scale. Although the scope of economics is broad, agricultural economics tends to focus on maximizing yields and efficiency at the farm level. Economics takes an anthropocentric approach to the value of the natural world: biodiversity, for example, is considered beneficial only to the extent that it is valued by people and increases profits. Some entities such as the European Union subsidize organic farming, in large part because these countries want to account for the externalities of reduced water use, reduced water contamination, reduced soil erosion, reduced carbon emissions, increased biodiversity, and assorted other benefits that result from organic farming.

Traditional organic farming is labor and knowledge-intensive whereas conventional farming is capital-intensive, requiring more energy and manufactured inputs.

Organic farmers in California have cited marketing as their greatest obstacle.

Geographic Producer Distribution

The markets for organic products are strongest in North America and Europe, which as of 2001 are estimated to have $6 and $8 billion respectively of the $20 billion global market. As of 2007 Australasia has 39% of the total organic farmland, including Australia's 1,180,000 hectares (2,900,000 acres) but 97 percent of this land is sprawling rangeland (2007). US sales are 20x as much. Europe farms 23 percent of global organic farmland (6,900,000 ha (17,000,000 acres)), followed.

by Latin America with 19 percent (5.8 million hectares - 14.3 million acres). Asia has 9.5 percent while North America has 7.2 percent. Africa has 3 percent.

Besides Australia, the countries with the most organic farmland are Argentina (3.1 million hectares - 7.7 million acres), China (2.3 million hectares - 5.7 million acres), and the United States (1.6 million hectares - 4 million acres). Much of Argentina's organic farmland is pasture, like that of Australia (2007). Spain, Germany, Brazil (the world's largest agricultural exporter), Uruguay, and the UK follow the United States in the amount of organic land (2007).

In the European Union (EU25) 3.9% of the total utilized agricultural area was used for organic production in 2005. The countries with the highest proportion of organic land were Austria (11%) and Italy (8.4%), followed by the Czech Republic and Greece (both 7.2%). The lowest figures were shown for Malta (0.1%), Poland (0.6%) and Ireland (0.8%). In 2009, the proportion of organic land in the EU grew to 4.7%. The countries with highest share of agricultural land were Liechtenstein (26.9%), Austria (18.5%) and Sweden (12.6%). 16% of all farmers in Austria produced organically in 2010. By the same year the proportion of organic land increased to 20%.: In 2005 168,000 ha (415,000 ac) of land in Poland was under organic management. In 2012, 288,261 hectares (712,308 acres) were under organic production, and there were about 15,500 organic farmers; retail sales of organic products were EUR 80 million in 2011. As of 2012 organic exports were part of the government's economic development strategy.

After the collapse of the Soviet Union in 1991, agricultural inputs that had previously been purchased from Eastern bloc countries were no longer available in Cuba, and many Cuban farms converted to organic methods out of necessity. Consequently, organic agriculture is a mainstream practice in Cuba, while it remains an alternative practice in most other countries. Cuba's organic strategy includes development of genetically modified crops; specifically corn that is resistant to the palomilla moth

Growth

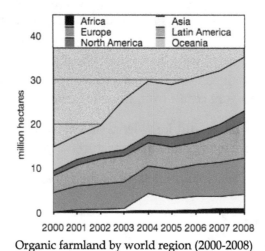

Organic farmland by world region (2000-2008)

In 2001, the global market value of certified organic products was estimated at USD $20 billion. By 2002, this was USD $23 billion and by 2007 more than USD $46 billion. By 2014, retail sales of organic products reached USD $80 billion worldwide. North America and Europe accounted for more than 90% of all organic product sales.

Organic agricultural land increased almost fourfold in 15 years, from 11 million hectares in 1999 to 43.7 million hectares in 2014. Between 2013 and 2014, organic agricultural land grew by 500,000 hectares worldwide, increasing in every region except Latin America. During this time period, Europe's organic farmland increased 260,000 hectares to 11.6 million total (+2.3%), Asia's increased 159,000 hectares to 3.6 million total (+4.7%), Africa's increased 54,000 hectares to 1.3 million total (+4.5%), and North America's increased 35,000 hectares to 3.1 million total (+1.1%). As of 2014, the country with the most organic land was Australia (17.2 million hectares), followed by Argentina (3.1 million hectares), and the United States (2.2 million hectares).

In 2013, the number of organic producers grew by almost 270,000, or more than 13%. By 2014, there were a reported 2.3 million organic producers in the world. Most of the total global increase took place in the Philippines, Peru, China, and Thailand. Overall, the majority of all organic producers are in India (650,000 in 2013), Uganda (190,552 in 2014), Mexico (169,703 in 2013) and the Philippines (165,974 in 2014).

Productivity

Studies comparing yields have had mixed results. These differences among findings can often be attributed to variations between study designs including differences in the crops studied and the methodology by which results were gathered.

A 2012 meta-analysis found that productivity is typically lower for organic farming than conventional farming, but that the size of the difference depends on context and in some cases may be very small. While organic yields can be lower than conventional yields, another meta-analysis published in Sustainable Agriculture Research in 2015, concluded that certain organic on-farm practices could help narrow this gap. Timely weed management and the application of manure in conjunction with legume forages/cover crops were shown to have positive results in increasing organic corn and soybean productivity. More experienced organic farmers were also found to have higher yields than other organic farmers who were just starting out.

Another meta-analysis published in the journal Agricultural Systems in 2011 analyzed 362 datasets and found that organic yields were on average 80% of conventional yields. The author's found that there are relative differences in this yield gap based on crop type with crops like soybeans and rice scoring higher than the 80% average and crops like wheat and potato scoring lower. Across global regions, Asia and Central Europe were found to have relatively higher yields and Northern Europe relatively lower than the average.

A 2007 study compiling research from 293 different comparisons into a single study to assess the overall efficiency of the two agricultural systems has concluded that "organic methods could produce enough food on a global per capita basis to sustain the current human population, and potentially an even larger population, without increasing the agricultural land base." The researchers also found that while in developed countries, organic systems on average produce 92% of the yield produced by conventional agriculture, organic systems produce 80% more than conventional farms in developing countries, because the materials needed for organic farming are more accessible than synthetic farming materials to farmers in some poor countries. This study was strongly contested by another study published in 2008, which stated, and was entitled, "Organic agriculture cannot feed the world" and said that the 2007 came up with "a major overestimation of the productivity of OA" "because data are misinterpreted and calculations accordingly are erroneous." Additional research needs to be conducted in the future to further clarify these claims.

Long Term Studies

A study published in 2005 compared conventional cropping, organic animal-based cropping, and organic legume-based cropping on a test farm at the Rodale Institute over 22 years. The study found that "the crop yields for corn and soybeans were similar in the organic animal, organic legume, and conventional farming systems". It also found that "significantly less fossil energy was expended to produce corn in the Rodale Institute's organic animal and organic legume systems than in the conventional production system. There was little difference in energy input between the different treatments for producing soybeans. In the organic systems, synthetic fertilizers and pesticides were generally not used". As of 2013 the Rodale study was ongoing and a thirty-year anniversary report was published by Rodale in 2012.

A long-term field study comparing organic/conventional agriculture carried out over 21 years in Switzerland concluded that "Crop yields of the organic systems averaged over 21 experimental years at 80% of the conventional ones. The fertilizer input, however, was 34 – 51% lower, indicating an efficient production. The organic farming systems used 20 – 56% less energy to produce a crop unit and per land area this difference was 36 – 53%. In spite of the considerably lower pesticide input the quality of organic products was hardly discernible from conventional analytically and even came off better in food preference trials and picture creating methods"

Profitability

In the United States, organic farming has been shown to be 2.9 to 3.8 times more profitable for the farmer than conventional farming when prevailing price premiums are taken into account. Globally, organic farming is between 22 and 35 percent more profitable for farmers than conventional methods, according to a 2015 meta-analysis of studies conducted across five continents.

The profitability of organic agriculture can be attributed to a number of factors. First, organic farmers do not rely on synthetic fertilizer and pesticide inputs, which can be costly. In addition, organic foods currently enjoy a price premium over conventionally produced foods, meaning that organic farmers can often get more for their yield.

The price premium for organic food is an important factor in the economic viability of organic farming. In 2013 there was a 100% price premium on organic vegetables and a 57% price premium for organic fruits. These percentages are based on wholesale fruit and vegetable prices, available through the United States Department of Agriculture's Economic Research Service. Price premiums exist not only for organic versus nonorganic crops, but may also vary depending on the venue where the product is sold: farmers markets, grocery stores, or wholesale to restaurants. For many producers, direct sales at farmers markets are most profitable because the farmer receives the entire markup, however this is also the most time and labor-intensive approach.

There have been signs of organic price premiums narrowing in recent years, which lowers the economic incentive for farmers to convert to or maintain organic production methods. Data from 22 years of experiments at the Rodale Institute found that, based on the current yields and production costs associated with organic farming in the United States, a price premium of only 10% is required to achieve parity with conventional farming. A separate study found that on a global scale, price premiums of only 5-7% percent were needed to break even with conventional methods. Without the price premium, profitability for farmers is mixed.

For markets and supermarkets organic food is profitable as well, and is generally sold at significantly higher prices than non-organic food.

Energy Efficiency

In the most recent assessments of the energy efficiency of organic versus conventional agriculture, results have been mixed regarding which form is more carbon efficient. Organic farm systems have more often than not been found to be more energy efficient, however, this is not always the case. More than anything, results tend to depend upon crop type and farm size.

A comprehensive comparison of energy efficiency in grain production, produce yield, and animal husbandry concluded that organic farming had a higher yield per unit of energy over the vast majority of the crops and livestock systems. For example, two studies - both comparing organically-versus conventionally-farmed apples - declare contradicting results, one saying organic farming is more energy efficient, the other saying conventionally is more efficient.

It has generally been found that the labor input per unit of yield was higher for organic systems compared with conventional production.

Sales and Marketing

Most sales are concentrated in developed nations. In 2008, 69% of Americans claimed to occasionally buy organic products, down from 73% in 2005. One theory for this change was that consumers were substituting "local" produce for "organic" produce.

Distributors

The USDA requires that distributors, manufacturers, and processors of organic products be certified by an accredited state or private agency. In 2007, there were 3,225 certified organic handlers, up from 2,790 in 2004.

Organic handlers are often small firms; 48% reported sales below $1 million annually, and 22% between $1 and $5 million per year. Smaller handlers are more likely to sell to independent natural grocery stores and natural product chains whereas large distributors more often market to natural product chains and conventional supermarkets, with a small group marketing to independent natural product stores. Some handlers work with conventional farmers to convert their land to organic with the knowledge that the farmer will have a secure sales outlet. This lowers the risk for the handler as well as the farmer. In 2004, 31% of handlers provided technical support on organic standards or production to their suppliers and 34% encouraged their suppliers to transition to organic. Smaller farms often join together in cooperatives to market their goods more effectively.

93% of organic sales are through conventional and natural food supermarkets and chains, while the remaining 7% of U.S. organic food sales occur through farmers' markets, foodservices, and other marketing channels.

Direct-to-Consumer Sales

In the 2012 Census, direct-to-consumer sales equaled $1.3 billion, up from $812 million in 2002, an increase of 60 percent. The number of farms that utilize direct-to-consumer sales was 144,530 in 2012 in comparison to 116,733 in 2002. Direct-to-consumer sales include farmers markets, community supported agriculture (CSA), on-farm stores, and roadside farm stands. Some organic farms also sell products direct to retailer, direct to restaurant and direct to institution. According to the 2008 Organic Production Survey, approximately 7% of organic farm sales went direct-to-consumers, 10% went direct to retailers, and approximately 83% went into wholesale markets. In comparison, only 0.4% of the value of convention agricultural commodities went direct-to-consumers.

While not all products sold at farmer's markets are certified organic, this direct-to-consumer avenue has become increasingly popular in local food distribution and has grown substantially since 1994. In 2014, there were 8,284 farmer's markets in comparison to 3,706 in 2004 and 1,755 in 1994, most of which are found in populated areas such as the Northeast, Midwest, and West Coast.

Labor and Employment

Organic production is more labor-intensive than conventional production. On the one hand, this increased labor cost is one factor that makes organic food more expensive. On the other hand, the increased need for labor may be seen as an "employment dividend" of organic farming, providing more jobs per unit area than conventional systems. The 2011 UNEP Green Economy Report suggests that "[a]n increase in investment in green agriculture is projected to lead to growth in employment of about 60 per cent compared with current levels" and that "green agriculture investments could create 47 million additional jobs compared with BAU2 over the next 40 years." The UNEP also argues that "[b]y greening agriculture and food distribution, more calories per person per day, more jobs and business opportunities especially in rural areas, and market-access opportunities, especially for developing countries, will be available."

World's Food Security

In 2007 the United Nations Food and Agriculture Organization (FAO) said that organic agriculture often leads to higher prices and hence a better income for farmers, so it should be promoted. However, FAO stressed that by organic farming one could not feed the current mankind, even less the bigger future population. Both data and models showed then that organic farming was far from sufficient. Therefore, chemical fertilizers were needed to avoid hunger. Other analysis by many agribusiness executives, agricultural and ecological scientists, and international agriculture experts revealed the opinion that organic farming would not only increase the world's food supply, but might be the only way to eradicate hunger.

FAO stressed that fertilizers and other chemical inputs can much increase the production, particularly in Africa where fertilizers are currently used 90% less than in Asia. For example, in Malawi the yield has been boosted using seeds and fertilizers. FAO also calls for using biotechnology, as it can help smallholder farmers to improve their income and food security.

Also NEPAD, development organization of African governments, announced that feeding Africans and preventing malnutrition requires fertilizers and enhanced seeds.

According to a more recent study in ScienceDigest, organic best management practices shows an average yield only 13% less than conventional. In the world's poorer nations where most of the world's hungry live, and where conventional agriculture's expensive inputs are not affordable by the majority of farmers, adopting organic management actually increases yields 93% on average, and could be an important part of increased food security.

Capacity Building in Developing Countries

Organic agriculture can contribute to ecologically sustainable, socio-economic development, especially in poorer countries. The application of organic principles enables employment of local resources (e.g., local seed varieties, manure, etc.) and therefore cost-effectiveness. Local and international markets for organic products show tremendous growth prospects and offer creative producers and exporters excellent opportunities to improve their income and living conditions.

Organic agriculture is knowledge intensive. Globally, capacity building efforts are underway, including localized training material, to limited effect. As of 2007, the International Federation of Organic Agriculture Movements hosted more than 170 free manuals and 75 training opportunities online.

In 2008 the United Nations Environmental Programme (UNEP) and the United Nations Conference on Trade and Development (UNCTAD) stated that "organic agriculture can be more conducive to food security in Africa than most conventional production systems, and that it is more likely to be sustainable in the long-term" and that "yields had more than doubled where organic, or near-organic practices had been used" and that soil fertility and drought resistance improved.

Millennium Development Goals

The value of organic agriculture (OA) in the achievement of the Millennium Development Goals (MDG), particularly in poverty reduction efforts in the face of climate change, is shown by its contribution to both income and non-income aspects of the MDGs. These benefits are expected to continue in the post-MDG era. A series of case studies conducted in selected areas in Asian countries by the Asian Development Bank Institute (ADBI) and published as a book compilation by ADB in Manila document these contributions to both income and non-income aspects of the MDGs. These include poverty alleviation by way of higher incomes, improved farmers' health owing to less chemical exposure, integration of sustainable principles into rural development policies, improvement of access to safe water and sanitation, and expansion of global partnership for development as small farmers are integrated in value chains.

A related ADBI study also sheds on the costs of OA programs and set them in the context of the costs of attaining the MDGs. The results show considerable variation across the case studies, suggesting that there is no clear structure to the costs of adopting OA. Costs depend on the efficiency of the OA adoption programs. The lowest cost programs were more than ten times less expensive than the highest cost ones. However, further analysis of the gains resulting from OA adoption reveals that the costs per person taken out of poverty was much lower than the estimates of the

World Bank, based on income growth in general or based on the detailed costs of meeting some of the more quantifiable MDGs (e.g., education, health, and environment).

Externalities

Agriculture imposes negative externalities (uncompensated costs) upon society through land and other resource use, biodiversity loss, erosion, pesticides, nutrient runoff, water usage, subsidy payments and assorted other problems. Positive externalities include self-reliance, entrepreneurship, respect for nature, and air quality. Organic methods reduce some of these costs. In 2000 uncompensated costs for 1996 reached 2,343 million British pounds or £208 per ha (£84.20/ac). A study of practices in the USA published in 2005 concluded that cropland costs the economy approximately 5 to 16 billion dollars ($30–96/ha - $12–39/ac), while livestock production costs 714 million dollars. Both studies recommended reducing externalities. The 2000 review included reported pesticide poisonings but did not include speculative chronic health effects of pesticides, and the 2004 review relied on a 1992 estimate of the total impact of pesticides.

It has been proposed that organic agriculture can reduce the level of some negative externalities from (conventional) agriculture. Whether the benefits are private or public depends upon the division of property rights.

Several surveys and studies have attempted to examine and compare conventional and organic systems of farming and have found that organic techniques, while not without harm, are less damaging than conventional ones because they reduce levels of biodiversity less than conventional systems do and use less energy and produce less waste when calculated per unit area.

A 2003 to 2005 investigation by the Cranfield University for the Department for Environment Food and Rural Affairs in the UK found that it is difficult to compare the Global Warming Potential (GWP), acidification and eutrophication emissions but "Organic production often results in increased burdens, from factors such as N leaching and N2O emissions", even though primary energy use was less for most organic products. N2O is always the largest GWP contributor except in tomatoes. However, "organic tomatoes always incur more burdens (except pesticide use)". Some emissions were lower "per area", but organic farming always required 65 to 200% more field area than non-organic farming. The numbers were highest for bread wheat (200+ % more) and potatoes (160% more).

The situation was shown dramatically in a comparison of a modern dairy farm in Wisconsin with one in New Zealand in which the animals grazed extensively. Using total farm emissions per kg milk produced as a parameter, the researchers showed that production of methane from belching was higher in the New Zealand farm, while carbon dioxide production was higher in the Wisconsin farm. Output of nitrous oxide, a gas with an estimated global warming potential 310 times that of carbon dioxide was also higher in the New Zealand farm. Methane from manure handling was similar in the two types of farm. The explanation for the finding relates to the different diets used on these farms, being based more completely on forage (and hence more fibrous) in New Zealand and containing less concentrate than in Wisconsin. Fibrous diets promote a higher proportion of acetate in the gut of ruminant animals, resulting in a higher production of methane that must be released by belching. When cattle are given a diet containing some concentrates (such as corn and soybean meal) in addition to grass and silage, the pattern of ruminal fermentation alters from acetate to mainly propionate. As a result, methane production is reduced. Capper et al. compared the environmental impact of US dairy production in 1944 and 2007. They calculated that the carbon "footprint" per billion kg (2.2 billion lb) of milk produced in 2007 was 37 percent that of equivalent milk production in 1944.

Environmental Impact and Emissions

Researchers at Oxford university analyzed 71 peer-reviewed studies and observed that organic products are sometimes worse for the environment. Organic milk, cereals, and pork generated higher greenhouse gas emissions per product than conventional ones but organic beef and olives had lower emissions in most studies. Usually organic products required less energy, but more land. Per unit of product, organic produce generates higher nitrogen leaching, nitrous oxide emissions, ammonia emissions, eutrophication and acidification potential than when conventionally grown. Other differences were not significant. The researchers concluded "Most of the studies that compared biodiversity in organic and conventional farming demonstrated lower environmental impacts from organic farming." The researchers believe that the ideal outcome would be to develop new systems that consider both the environment, including setting land aside for wildlife and sustainable forestry, and the development of ways to produce the highest yields possible using both conventional and organic methods.

Proponents of organic farming have claimed that organic agriculture emphasizes closed nutrient cycles, biodiversity, and effective soil management providing the capacity to mitigate and even reverse the effects of climate change and that organic agriculture can decrease fossil fuel emissions. "The carbon sequestration efficiency of organic systems in temperate climates is almost double (575-700 kg carbon per ha per year - 510-625 lb/ac/an) that of conventional treatment of soils, mainly owing to the use of grass clovers for feed and of cover crops in organic rotations."

Critics of organic farming methods believe that the increased land needed to farm organic food could potentially destroy the rainforests and wipe out many ecosystems.

Nutrient Leaching

According to the meta-analysis of 71 studies, nitrogen leaching, nitrous oxide emissions, ammonia emissions, eutrophication potential and acidification potential were higher for organic products, although in one study "nitrate leaching was 4.4-5.6 times higher in conventional plots than organic plots".

Excess nutrients in lakes, rivers, and groundwater can cause algal blooms, eutrophication, and subsequent dead zones. In addition, nitrates are harmful to aquatic organisms by themselves.

Land Use

The Oxford meta-analysis of 71 studies proved that organic farming requires 84% more land, mainly due to lack of nutrients but sometimes due to weeds, diseases or pests, lower yielding animals and land required for fertility building crops. While organic farming does not necessarily save land for wildlife habitats and forestry in all cases, the most modern breakthroughs in organic are addressing these issues with success.

Professor Wolfgang Branscheid says that organic animal production is not good for the environment, because organic chicken requires doubly as much land as "conventional" chicken and organic pork a quarter more. According to a calculation by Hudson Institute, organic beef requires triply as much land. On the other hand, certain organic methods of animal husbandry have been shown to restore desertified, marginal, and/or otherwise unavailable land to agricultural productivity and wildlife. Or by getting both forage and cash crop production from the same fields simultaneously, reduce net land use.

In England organic farming yields 55% of normal yields. While in other regions of the world, organic methods have started producing record yields.

Pesticides

A sign outside of an organic apple orchard in Pateros, Washington reminding orchardists
not to spray pesticides on these trees.

Food Quality and Safety

While there may be some differences in the amounts of nutrients and anti-nutrients when organically produced food and conventionally produced food are compared, the variable nature of food production and handling makes it difficult to generalize results, and there is insufficient evidence to make claims that organic food is safer or healthier than conventional food. Claims that organic food tastes better are not supported by evidence.

Soil Conservation

Supporters claim that organically managed soil has a higher quality and higher water retention. This may help increase yields for organic farms in drought years. Organic farming can build up soil organic matter better than conventional no-till farming, which suggests long-term yield benefits from organic farming. An 18-year study of organic methods on nutrient-depleted soil concluded that conventional methods were superior for soil fertility and yield for nutrient-depleted soils in cold-temperate climates, arguing that much of the benefit from organic farming derives from imported materials that could not be regarded as self-sustaining.

In *Dirt: The Erosion of Civilizations*, geomorphologist David Montgomery outlines a coming crisis from soil erosion. Agriculture relies on roughly one meter of topsoil, and that is being depleted ten times faster than it is being replaced. No-till farming, which some claim depends upon pesticides, is one way to minimize erosion. However, a 2007 study by the USDA's Agricultural Research Service has found that manure applications in tilled organic farming are better at building up the soil than no-till.

Biodiversity

The conservation of natural resources and biodiversity is a core principle of organic production. Three broad management practices (prohibition/reduced use of chemical pesticides and inorganic fertilizers; sympathetic management of non-cropped habitats; and preservation of mixed farming) that are largely intrinsic (but not exclusive) to organic farming are particularly beneficial for farmland wildlife. Using practices that attract or introduce beneficial insects, provide habitat for birds and mammals, and provide conditions that increase soil biotic diversity serve to supply vital ecological services to organic production systems. Advantages to certified organic operations that implement these types of production practices include: 1) decreased dependence on outside fertility inputs; 2) reduced pest management costs; 3) more reliable sources of clean water; and 4) better pollination.

Nearly all non-crop, naturally occurring species observed in comparative farm land practice studies show a preference for organic farming both by abundance and diversity. An average of 30% more species inhabit organic farms. Birds, butterflies, soil microbes, beetles, earthworms, spiders, vegetation, and mammals are particularly affected. Lack of herbicides and pesticides improve biodiversity fitness and population density. Many weed species attract beneficial insects that improve soil qualities and forage on weed pests. Soil-bound organisms often benefit because of increased bacteria populations due to natural fertilizer such as manure, while experiencing reduced intake of herbicides and pesticides. Increased biodiversity, especially from beneficial soil microbes and mycorrhizae have been proposed as an explanation for the high yields experienced by some organic plots, especially in light of the differences seen in a 21-year comparison of organic and control fields.

Biodiversity from organic farming provides capital to humans. Species found in organic farms enhance sustainability by reducing human input (e.g., fertilizers, pesticides).

The USDA's Agricultural Marketing Service (AMS) published a *Federal Register* notice on 15 January 2016, announcing the National Organic Program (NOP) final guidance on Natural Resources and Biodiversity Conservation for Certified Organic Operations. Given the broad scope of natural resources which includes soil, water, wetland, woodland and wildlife, the guidance provides examples of practices that support the underlying conservation principles and demonstrate compliance with USDA organic regulations § 205.200. The final guidance provides organic certifiers and farms with examples of production practices that support conservation principles and comply with the USDA organic regulations, which require operations to maintain or improve natural resources. The final guidance also clarifies the role of certified operations (to submit an OSP to a certifier), certifiers (ensure that the OSP describes or lists practices that explain the operator's monitoring plan and practices to support natural resources and biodiversity conservation), and inspectors (onsite inspection) in the implementation and verification of these production practices.

A wide range of organisms benefit from organic farming, but it is unclear whether organic methods confer greater benefits than conventional integrated agri-environmental programs. Organic farming is often presented as a more biodiversity-friendly practice, but the generality of the beneficial effects of organic farming is debated as the effects appear often species- and context-dependent, and current research has highlighted the need to quantify the relative effects of local- and landscape-scale management on farmland biodiversity. There are four key issues when comparing the impacts on biodiversity of organic and conventional farming: (1) It remains unclear whether a holistic whole-farm approach (i.e. organic) provides greater benefits to biodiversity than carefully targeted prescriptions applied to relatively small areas of cropped and/or non-cropped habitats within conventional agriculture (i.e. agri-environment schemes); (2) Many comparative studies encounter methodological problems, limiting their ability to draw quantitative conclusions; (3) Our knowledge of the impacts of organic farming in pastoral and upland agriculture is limited; (4) There remains a pressing need for longitudinal, system-level studies in order to address these issues and to fill in the gaps in our knowledge of the impacts of organic farming, before a full appraisal of its potential role in biodiversity conservation in agroecosystems can be made.

Regional Support for Organic Farming

India

In India, states such as Sikkim and Kerala have planned to shift to fully organic cultivation by 2015 and 2016 respectively.

Principle of Organic Farming

Organic Horticulture

An organic garden on a school campus

Organic horticulture is the science and art of growing fruits, vegetables, flowers, or ornamental plants by following the essential principles of organic agriculture in soil building and conservation, pest management, and heirloom variety preservation.

The Latin words *hortus* (garden plant) and *cultura* (culture) together form *horticulture*, classically defined as the culture or growing of garden plants. *Horticulture* is also sometimes defined simply as "agriculture minus the plough." Instead of the plough, horticulture makes use of human labour and gardener's hand tools, although some small machine tools like rotary tillers are commonly employed now.

General

Mulches, cover crops, compost, manures, vermicompost, and mineral supplements are soil-building mainstays that distinguish this type of farming from its commercial counterpart. Through attention to good healthy soil condition, it is expected that insect, fungal, or other problems that sometimes plague plants can be minimized. However, pheromone traps, insecticidal soap sprays, and other pest-control methods available to organic farmers are also utilized by organic horticulturists.

Horticulture involves five areas of study. These areas are floriculture (includes production and marketing of floral crops), landscape horticulture (includes production, marketing and maintenance of landscape plants), olericulture (includes production and marketing of vegetables), pomology (includes production and marketing of fruits), and postharvest physiology (involves maintaining quality and preventing spoilage of horticultural crops). All of these can be, and sometimes are, pursued according to the principles of organic cultivation.

Organic horticulture (or organic gardening) is based on knowledge and techniques gathered over thousands of years. In general terms, organic horticulture involves natural processes, often taking place over extended periods of time, and a sustainable, holistic approach - while chemical-based horticulture focuses on immediate, isolated effects and reductionist strategies.

Organic gardening systems

There are a number of formal organic gardening and farming systems that prescribe specific techniques. They tend to be more specific than, and fit within, general organic standards. Forest gardening, a fully organic food production system which dates from prehistoric times, is thought to be the world's oldest and most resilient agroecosystem.

Biodynamic farming is an approach based on the esoteric teachings of Rudolf Steiner. The Japanese farmer and writer Masanobu Fukuoka invented a no-till system for small-scale grain production that he called Natural Farming. French intensive gardening and biointensive methods and SPIN Farming (Small Plot INtensive) are all small scale gardening techniques. These techniques were brought to the United States by Alan Chadwick in the 1930s. This method has since been promoted by John Jeavons, Director of Ecology Action. A garden is more than just a means of providing food, it is a model of what is possible in a community - everyone could have a garden of some kind (container, growing box, raised bed) and produce healthy, nutritious organic food, a farmers market, a place to pass on gardening experience, and a sharing of bounty, promoting a more sustainable way of living that would encourage their local economy. A simple 4' x 8' (32 square feet) raised bed garden based on the principles of bio-intensive planting and square foot gardening uses fewer nutrients and less water, and could keep a family, or community, supplied with an abundance of healthy, nutritious organic greens, while promoting a more sustainable way of living.

Organic gardening is designed to work with the ecological systems and minimally disturb the Earth's natural balance. Because of this organic farmers have been interested in reduced-tillage methods. Conventional agriculture uses mechanical tillage, which is plowing or sowing, which is harmful to the environment. The impact of tilling in organic farming is much less of an issue. Ploughing speeds up erosion because the soil remains uncovered for a long period of time and if it has a low content of organic matter the structural stability of the soil decreases. Organic farmers use techniques such as mulching, planting cover crops, and intercropping, to maintain a soil cover throughout most of the year. The use of compost, manure mulch and other organic fertilizers yields a higher organic content of soils on organic farms and helps limit soil degradation and erosion.

Other methods can also be used to supplement an existing garden. Methods such as composting, or vermicomposting. These practices are ways of recycling organic matter into some of the best organic fertilizers and soil conditioner. Vermicompost is especially easy. The byproduct is also an excellent source of nutrients for an organic garden.

Pest control approaches

Differing approaches to pest control are equally notable. In chemical horticulture, a specific insecticide may be applied to quickly kill off a particular insect pest. Chemical controls can dramatically reduce pest populations in the short term, yet by unavoidably killing (or starving) natural control insects and animals, cause an increase in the pest population in the long term, thereby creating an ever increasing problem. Repeated use of insecticides and herbicides also encourages rapid natural selection of resistant insects, plants and other organisms, necessitating increased use, or requiring new, more powerful controls.

In contrast, organic horticulture tends to tolerate some pest populations while taking the long view. Organic pest control requires a thorough understanding of pest life cycles and interactions, and involves the cumulative effect of many techniques, including:

- Allowing for an acceptable level of pest damage
- Encouraging predatory beneficial insects to flourish and eat pests
- Encouraging beneficial microorganisms
- Careful plant selection, choosing disease-resistant varieties
- Planting companion crops that discourage or divert pests
- Using row covers to protect crop plants during pest migration periods
- Rotating crops to different locations from year to year to interrupt pest reproduction cycles
- Using insect traps to monitor and control insect populations

Each of these techniques also provides other benefits, such as soil protection and improvement, fertilization, pollination, water conservation and season extension. These benefits are both complementary and cumulative in overall effect on site health. Organic pest control and biological pest control can be used as part of integrated pest management (IPM). However, IPM can include the use of chemical pesticides that are not part of organic or biological techniques.

Impact on the Global Food Supply

One controversy associated with organic food production is the matter of food produced per acre. Even with good organic practices, organic agriculture may be five to twenty-five percent less productive than conventional agriculture, depending on the crop.

Much of the productivity advantage of conventional agriculture is associated with the use of nitrogen fertilizer. However, the use, and especially the overuse, of nitrogen fertilizer has negative effects such as nitrogen runoff harming natural water supplies and increased global warming.

Organic methods have other advantages, such as healthier soil, that may make organic farming more resilient, and therefore more reliable in producing food, in the face of challenges such as climate change.

As well, world hunger is not primarily an issue of agricultural yields, but distribution and waste.

Extensive Farming

A small farm in the Swiss mountains. The land here is mostly rock and the slopes are very steep – likely unusable for agriculture, but can provide productive conditions for pigs

Extensive farming or extensive agriculture (as opposed to intensive farming) is an agricultural production system that uses small inputs of labor, fertilizers, and capital, relative to the land area being farmed.

Extensive farming most commonly refers to sheep and cattle farming in areas with low agricultural productivity, but can also refer to large-scale growing of wheat, barley, cooking oils and other grain crops in areas like the Murray-Darling Basin. Here, owing to the extreme age and poverty of the soils, yields per hectare are very low, but the flat terrain and very large farm sizes mean yields per unit of labour are high. Nomadic herding is an extreme example of extensive farming, where herders move their animals to use feed from occasional rainfalls.

Geography

Extensive farming is found in the mid-latitude sections of most continents, as well as in desert regions where water for cropping is not available. The nature of extensive farming means it requires less rainfall than intensive farming. The farm is usually large in comparison with the numbers working and money spent on it. In most parts of Western Australia, pastures are so poor that only one sheep to the square mile can be supported

Just as the demand has led to the basic division of cropping and pastoral activities, these areas can also be subdivided depending on the region's rainfall, vegetation type and agricultural activity within the area and the many other parentheses related to this data.

Advantages

Extensive farming has a number of advantages over intensive farming:

1. Less labour per unit areas is required to farm large areas, especially since expensive alterations to land (like terracing) are completely absent.

2. Mechanisation can be used more effectively over large, flat areas.

3. Greater efficiency of labour means generally lower product prices.

4. Animal welfare is generally improved because animals are not kept in stifling conditions.

5. Lower requirements of inputs such as fertilizers.

6. If animals are grazed on pastures native to the locality, there is less likely to be problems with exotic species.

7. Local environment and soil are not damaged by overuse of chemicals.

8. The use of machinery and scientific methods of farming produce a large quantity of crops.

Disadvantages

Extensive farming can have the following problems:

1. Yields tend to be much lower than with intensive farming in the short term.

2. Large land requirements limit the habitat of wild species (in some cases, even very low stocking rates can be dangerous), as is the case with intensive farming

Rainfed Agriculture

The term rainfed agriculture is used to describe farming practises that rely on rainfall for water. It provides much of the food consumed by poor communities in developing countries. For example, rainfed agriculture accounts for more than 95% of farmed land in sub-Saharan Africa, 90% in Latin America, 75% in the Near East and North Africa; 65% in East Asia and 60% in South Asia.

Levels of productivity, particularly in parts of sub-Saharan Africa and South Asia, are low due to degraded soils, high levels of evaporation, droughts, floods and a general lack of effective water management. A major study into water use by agriculture, known as the Comprehensive Assessment of Water Management in Agriculture, coordinated by the International Water Management Institute, noted a close correlation between hunger, poverty and water. However, it concluded that there was much opportunity to raise productivity from rainfed farming.

The authors considered that managing rainwater and soil moisture more effectively, and using supplemental and small-scale irrigation, held the key to helping the greatest number of poor people. It called for a new era of water investments and policies for upgrading rainfed agriculture that would go beyond controlling field-level soil and water to bring new freshwater sources through better local management of rainfall and runoff.

The importance of rainfed agriculture varies regionally but produces most food for poor communities in developing countries. In subSaharan Africa more than 95% of the farmed land is rainfed, while the corresponding figure for Latin America is almost 90%, for South Asia about 60%, for East Asia 65% and for the Near East and North Africa 75%. Most countries in the world depend primarily on rainfed agriculture for their grain food. Despite large strides made in improving productivity and environmental conditions in many developing countries, a great number of poor families in Africa and Asia still face poverty, hunger, food insecurity and malnutrition where rainfed agriculture is the main agricultural activity. These problems are exacerbated by adverse biophysical growing conditions and the poor socioeconomic infrastructure in many areas in the semi-arid tropics (SAT). The SAT is the home to 38% of the developing countries' poor, 75% of whom live in rural areas. Over 45% of the world's hungry and more than 70% of its malnourished children live in the SAT

There is a correlation between poverty, hunger and water stress. The UN Millennium Development Project has identified the 'hot spot' countries in the world suffering from the largest prevalence of malnourishment. These countries coincide closely with those located in the semi-arid and dry subhumid hydroclimates in the world, i.e. savannahs and steppe ecosystems, where rainfed agriculture is the dominating source of food and where water constitutes a key limiting factor to crop growth. Of the 850 million undernourished people in the world, essentially all live in poor, developing countries, which predominantly are located in tropical regions.

Since the late 1960s, agricultural land use has expanded by 20–25%, which has contributed to approximately 30% of the overall grain production growth during the period. The remaining yield outputs originated from intensification through yield increases per unit land area. However, the regional variation is large, as is the difference between irrigated and rainfed agriculture. In developing countries rainfed grain yields are on average 1.5 t/ha, compared with 3.1 t/ha for irrigated yields (Rosegrant et al., 2002), and increase in production from rainfed agriculture has mainly originated from land expansion. Trends are clearly different for different regions. With 99% rainfed production of main cereals such as maize, millet and sorghum, the cultivated cereal area in sub-Saharan Africa has doubled since 1960 while the yield per unit of land has been nearly stagnant for these staple crops (FAOSTAT, 2005). In South Asia, there has been a major shift away from more drought-tolerant, low-yielding crops such as sorghum and millet, while wheat and maize hasapproximately doubled in area since 1961 (FAOSTAT, 2005). During the same period, the yield per unit of land for maize and wheat has more than doubled. For predominantly rainfed systems, maize crops per unit of land have nearly tripled and wheat more than doubled during the same time period. Rainfed maize yield differs substantially between regions. In Latin America (including the Caribbean) it exceeds 3 t/ha, while in South Asia it is around 2 t/ha and in subSaharan Africa it only just exceeds 1 t/ha. This can be compared with maize yields in the USA or southern Europe, which normally amount to approximately 7–10 t/ha (most maize in these regions is irrigated). The average regional yield per unit of land for wheat in Latin America (including the Caribbean) and South Asia is similar to the average yield output of 2.5–2.7 t/ha in North America. In comparison, wheat yield in Western Europe is approximately twice as large (5 t/ha), while in sub-Saharan Africa it remains below 2 t/ha. In view of the historic regional difference in development of yields, there appears to exist a significant potential for raised yields in rainfed agriculture, particularly in sub-Saharan Africa and South Asia.

Rural development through sustainable management of land and water resources gives a plausible solution for alleviating rural poverty and improving the livelihoods of the rural poor. In an effective convergence mode for improving the rural livelihoods in the target districts, with watersheds as the operational units, a holistic integrated systems approach by drawing attention to the past experiences, existing opportunities and skills, and supported partnerships can enable change and improve the livelihoods of the rural poor. The well-being of the rural poor depends on fostering their fair and equitable access to productive resources. The rationale behind convergence through watersheds has been that these watersheds help in 'cross-learning' and drawing on a wide range of experiences from different sectors. A significant conclusion is that there should be a balance between attending to needs and priorities of rural livelihoods and enhancing positive directions of change by building effective and sustainable partnerships. Based on the experience and performance of the existing integrated community watersheds in different socioeconomic environments, appropriate exit strategies, which include proper sequencing of interventions, building up of financial, technical and organizational capacity of local communities to internalize and sustain interventions, and the requirement for any minimal external technical and organizational support needs to be identified.

While absolute grain yield variations exist between different global locations as cited in this article, the potential for improved rain fed grain yields may be less than is suggested by a comparison between sub-Saharan and European locations for example, this applies particularly in areas where grain yield is primarily determined by the growing season rainfall. A more accurate formula for measuring yield potential is y x a = X, where X is grain yield in Kg/hectare, y is millimetres of growing season rainfall and a is the variable yield factor, a number that may vary somewhere between

5 and 20. Thus assuming a variable yield factor of 15 and a growing season rainfall of 220mm the formula would express as 220mm x 15 = 3300 kg/ha yield potential. This formula, while not taking account of either the carryover benefits of stored rainfall in the soil profile resulting from out of season rainfall or the impact of temperature or soil fertility, still gives a more accurate picture of the degree to which actual grain yields are matching the region's potential yields and is a better basis for comparison between very different regions such as Europe and the sub-Sahara. A European yield of 5000 kg/ha from a rainfall of 500mm results in a variable yield factor of 10 while a yield of 2500 kg/ha in a 200mm rainfall area has a variable yield factor of 12.5, in such an example the lower yielding crop has actually been 25% more efficient in its rainfall utilisation than the higher yielding crop. Agronomy needs to be based on attaining the highest possible variable yield factor rather than the highest absolute yield, factor numbers as high as 17+ are achievable.

Biodynamic Agriculture

Biodynamic agriculture is a form of alternative agriculture very similar to organic farming, but which includes various esoteric concepts drawn from the ideas of Rudolf Steiner (1861–1925). Initially developed in the 1920s, it was the first of the organic agriculture movements. It treats soil fertility, plant growth, and livestock care as ecologically interrelated tasks, emphasizing spiritual and mystical perspectives.

Biodynamics has much in common with other organic approaches – it emphasizes the use of manures and composts and excludes the use of artificial chemicals on soil and plants. Methods unique to the biodynamic approach include its treatment of animals, crops, and soil as a single system; an emphasis from its beginnings on local production and distribution systems; its use of traditional and development of new local breeds and varieties; and the use of an astrological sowing and planting calendar. Biodynamic agriculture uses various herbal and mineral additives for compost additives and field sprays; these are sometimes prepared by controversial methods, such as burying ground quartz stuffed into the horn of a cow, which are said to harvest "cosmic forces in the soil", that are more akin to sympathetic magic than agronomy.

As of 2016 biodynamic techniques were used on 161,074 hectares in 60 countries. Germany accounts for 45% of the global total; the remainder average 1750 ha per country. Biodynamic methods of cultivating grapevines have been taken up by several notable vineyards. There are certification agencies for biodynamic products, most of which are members of the international biodynamics standards group Demeter International.

No difference in beneficial outcomes has been scientifically established between certified biodynamic agricultural techniques and similar organic and integrated farming practices. Critics have characterized biodynamic agriculture as pseudoscience on the basis of a lack of strong evidence for its efficacy and skepticism about aspects characterized as magical thinking.

History

Biodynamics was the first modern organic agriculture. Its development began in 1924 with a series of eight lectures on agriculture given by philosopher Rudolf Steiner at Schloss Koberwitz in Silesia, Germany, (now Kobierzyce in Poland southwest of Wrocław). These lectures, the first known presentation of organic agriculture, were held in response to a request by farmers who noticed degraded soil conditions and a deterioration in the health and quality of crops and livestock resulting from the use of chemical fertilizers. The one hundred eleven attendees, less than half of whom were farmers, came from six countries, primarily Germany and Poland. The lectures were published in November 1924; the first English translation appeared in 1928 as *The Agriculture Course*.

Steiner emphasized that the methods he proposed should be tested experimentally. For this purpose, Steiner established a research group, the "Agricultural Experimental Circle of Anthroposophical Farmers and Gardeners of the General Anthroposophical Society". This research group attracted, in the interval 1924 to 1939, about 800 members from around the world, including Europe, the Americas and Australasia. Another group, the "Association for Research in Anthroposophical Agriculture" (Versuchsring anthroposophischer Landwirte), directed by the German agronomist Erhard Bartsch, was formed to test the effects of biodynamic methods on the life and health of soil, plants and animals; the group published a monthly journal *Demeter*. Bartsch was also instrumental in developing a sales organisation for biodynamic products, Demeter, which still exists today. The Research Association was renamed The Imperial Association for Biodynamic Agriculture (Reichsverband für biologisch-dynamische Wirtschaftsweise) in 1933. It was dissolved by the National Socialist regime in 1941. In 1931 the association had 250 members in Germany, 109 in Switzerland, 104 in other European countries and 24 outside Europe. The oldest biodynamic farms are the Wurzerhof in Austria and Marienhöhe in Germany.

In 1938, Ehrenfried Pfeiffer's text *Bio-Dynamic Farming and Gardening* was published in five languages – English, Dutch, Italian, French, and German; this became the standard work in the field for several decades. In July 1939, at the invitation of Walter James, 4th Baron Northbourne, Pfeiffer travelled to the UK and presented the Betteshanger Summer School and Conference on Biodynamic Farming' at Northbourne's farm in Kent. The conference has been described as the 'missing link' between biodynamic agriculture and organic farming because, in the year after Betteshanger, Northbourne published his manifesto of organic farming, *Look to the Land*, in which he coined the term 'organic farming' and praised the methods of Rudolf Steiner. In the 1950s, Hans Mueller was encouraged by Steiner's work to create the organic-biological farming method in Switzerland; this later developed to become the largest certifier of organic products in Europe, *Bioland*.

Today biodynamics is practiced in more than 50 countries worldwide and in a variety of circumstances, ranging from temperate arable farming, viticulture in France, cotton production in Egypt, to silkworm breeding in China. Germany accounts for nearly half of the world's biodynamic agriculture. Demeter International is the primary certification agency for farms and gardens using the methods.

Geographic Developments

- In Australia, the first biodynamic farmer was Ernesto Genoni who in 1928 joined the Experimental Circle of Anthroposophical Farmers and Gardeners, followed soon after by his brother Emilio Genoni. Bob Williams presented the first public lecture in Australia on biodynamic agriculture on 26 June 1938 at the home of the architects Walter Burley Griffin and Marion Mahony Griffin at Castlecrag, Sydney. Since the 1950s research work has continued at the Biodynamic Research Institute (BDRI) in Powelltown, near Melbourne under the direction of Alex Podolinsky. In 1989 Biodynamic Agriculture Australia was established, as a not for profit association.

- In 1928 the *Anthroposophical Agricultural Foundation* was founded in England; this is now called the *Biodynamic Agriculture Association*. In 1939, Britain's first biodynamic agriculture conference, the Betteshanger Summer School and Conference on Biodynamic Agriculture, was held at Lord Northbourne's farm in Kent; Ehrenfried Pfeiffer was the lead presenter.

- In the United States, the Biodynamic Farming & Gardening Association was founded in 1938 as a New York state corporation.

- In France the International Federation of Organic Agriculture Movements (IFOAM) was formed in 1972 with five founding members, one of which was the Swedish Biodynamic Association.

- The University of Kassel had a Department of Biodynamic Agriculture from 2006 to March 2011.

Biodynamic Method of Farming

In common with other forms of organic agriculture, biodynamic agriculture uses management practices that are intended to "restore, maintain and enhance ecological harmony." Central features include crop diversification, the avoidance of chemical soil treatments and off-farm inputs generally, decentralized production and distribution, and the consideration of celestial and terrestrial influences on biological organisms. The Demeter Association recommends that "(a) minimum of ten percent of the total farm acreage be set aside as a biodiversity preserve. That may include but is not limited to forests, wetlands, riparian corridors, and intentionally planted insectaries. Diversity in crop rotation and perennial planting is required: no annual crop can be planted in the same field for more than two years in succession. Bare tillage year round is prohibited so land needs to maintain adequate green cover."

The Demeter Association also recommends that the individual design of the land "by the farmer, as determined by site conditions, is one of the basic tenets of biodynamic agriculture. This principle emphasizes that humans have a responsibility for the development of their ecological and social environment which goes beyond economic aims and the principles of descriptive ecology." Crops, livestock, and farmer, and "the entire socioeconomic environment" form a unique interaction, which biodynamic farming tries to "actively shape ...through a variety of management practices. The prime objective is always to encourage healthy conditions for life": soil fertility, plant and animal health, and product quality. "The farmer seeks to enhance and support the forces of nature that lead to healthy crops, and rejects farm management practices that damage the environment, soil plant, animal or human health....the farm is conceived of as an organism, a self-contained entity with its own individuality," holistically conceived and self-sustaining. "Disease and insect control are addressed through botanical species diversity, predator habitat, balanced crop nutrition, and attention to light penetration and airflow. Weed control emphasizes prevention, including timing of planting, mulching, and identifying and avoiding the spread of invasive weed species."

Biodynamic agriculture differs from many forms of organic agriculture in its spiritual, mystical, and astrological orientation. It shares a spiritual focus, as well as its view toward improving humanity, with the "nature farming" movement in Japan. Important features include the use of livestock manures to sustain plant growth (recycling of nutrients), maintenance and improvement of soil quality, and the health and well being of crops and animals. Cover crops, green manures and crop rotations are used extensively and the farms to foster the diversity of plant and animal life, and to enhance the biological cycles and the biological activity of the soil.

Biodynamic farms often have a cultural component and encourage local community, both through developing local sales and through on-farm community building activities. Some biodynamic farms use the Community Supported Agriculture model, which has connections with social threefolding.

Compared to non-organic agriculture, BD farming practices have been found to be more resilient to environmental challenges, to foster a diverse biosphere, and to be more energy efficient, factors Eric Lichtfouse describes being of increasing importance in the face of climate change, energy scarcity and population growth.

Biodynamic Preparations

In his "agricultural course" Steiner prescribed nine different preparations to aid fertilization, and described how these were to be prepared. Steiner believed that these preparations mediated terrestrial and cosmic forces into the soil. The prepared substances are numbered 500 through 508, where the first two are used for preparing fields, and seven are used for making compost. A long term trial (DOK experiment) evaluating the biodynamic farming system in comparison with organic and conventional farming systems, found that both organic farming and biodynamic farming resulted in enhanced soil properties, but had lower yields than conventional farming. Regarding compost development beyond accelerating the initial phase of composting, some positive effects have been noted:

- The field sprays contain substances that stimulate plant growth include cytokinins.

- Some improvement in nutrient content of compost.

Although the preparations have direct nutrient values, their purpose in biodynamics is to support the self-regulating capacities of the soil biota in the case of 500 and 501 and the biological life resident in the composting organics, as well as the mature compost itself, in the others.

Field Preparations

Field preparations, for stimulating humus formation:

- 500: (horn-manure) a humus mixture prepared by filling the horn of a cow with cow manure and burying it in the ground (40–60 cm below the surface) in the autumn. It is left to decompose during the winter and recovered for use the following spring.

- 501: Crushed powdered quartz prepared by stuffing it into a horn of a cow and buried into the ground in spring and taken out in autumn. It can be mixed with 500 but usually prepared on its own (mixture of 1 tablespoon of quartz powder to 250 liters of water) The mixture is sprayed under very low pressure over the crop during the wet season, in an attempt to prevent fungal diseases. It should be sprayed on an overcast day or early in the morning to prevent burning of the leaves.

The application rate of the biodynamic field spray preparations (i.e., **500** and **501**) are 300 grams per hectare of horn manure and 5 grams per hectare of horn silica. These are made by stirring the ingredients into 20-50 liters of water per hectare for an hour, using a prescribed method.

Compost Preparations

Compost preparations, used for preparing compost, employ herbs which are frequently used in medicinal remedies. Many of the same herbs are used in organic practices to make foliar fertilizers, turned into the soil as green manure, or in composting. The preparations include:

- 502: Yarrow blossoms (*Achillea millefolium*) are stuffed into urinary bladders from Red Deer (*Cervus elaphus*), placed in the sun during summer, buried in earth during winter and retrieved in the spring.

- 503: Chamomile blossoms (*Matricaria recutita*) are stuffed into small intestines from cattle buried in humus-rich earth in the autumn and retrieved in the spring.

- 504: Stinging nettle (*Urtica dioica*) plants in full bloom are stuffed together underground surrounded on all sides by peat for a year.

- 505: Oak bark (*Quercus robur*) is chopped in small pieces, placed inside the skull of a domesticated animal, surrounded by peat and buried in earth in a place where lots of rain water runs past.

- 506: Dandelion flowers (*Taraxacum officinale*) are stuffed into the mesentery of a cow and buried in earth during winter and retrieved in the spring.

- 507: Valerian flowers (*Valeriana officinalis*) are extracted into water.

- 508: Horsetail (*Equisetum*).

The compost preparations are applied with quantities of 1–2 cm^3 each per 10 m^3 compost, farmyard manure or liquid manure. The preparations should then be evenly sprayed out on the land as soon as possible after stirring.

One to three grams (a teaspoon) of each preparation is added to a dung heap by digging 50 cm deep holes with a distance of 2 meters from each other, except for the 507 preparation, which is stirred into 5 liters of water and sprayed over the entire compost surface. All preparations are thus used in homeopathic quantities. Each compost preparation is designed to guide a particular decomposition process in the composting mass. One study found that the oak bark preparation improved disease resistance in zucchini.

Planting Calendar

The approach considers that there are lunar and astrological influences on soil and plant development—for example, choosing to plant, cultivate or harvest various crops based on both the phase of the moon and the zodiacal constellation the moon is passing through, and also depending on whether the crop is the root, leaf, flower, or fruit of the plant. This aspect of biodynamics has been termed "astrological" in nature.

Seed Production

Biodynamic agriculture has focused on the open pollination of seeds (with farmers thereby generally growing their own seed) and the development of locally adapted varieties. The seed stock is not controlled by large, multinational seed companies.

Biodynamic Certification

The biodynamic certification Demeter, created in 1924, was the first certification and labelling system for organic production. To receive certification as a biodynamic farm, the farm must meet the following standards: agronomic guidelines, greenhouse management, structural components, livestock guidelines, and post harvest handling and processing procedures.

The term *Biodynamic* is a trademark held by the Demeter association of biodynamic farmers for the purpose of maintaining production standards used both in farming and processing foodstuffs.(This is not a trademark held privately in New Zealand) The trademark is intended to protect both the consumer and the producers of biodynamic produce. Demeter International an organization of member countries; each country has its own Demeter organization which is required to meet international production standards (but can also exceed them). The original Demeter organization was founded in 1928; the U.S. Demeter Association was formed in the 1980s and certified its first farm in 1982. In France, Biodivin certifies biodynamic wine. In Egypt, SEKEM has created the Egyptian Biodynamic Association (EBDA), an association that provides training for farmers to become certified. As of 2006, more than 200 wineries worldwide were certified as biodynamic; numerous other wineries employ biodynamic methods to a greater or lesser extent.

Effectiveness

Research into biodynamic farming has been complicated by the difficulty of isolating the distinctively biodynamic aspects when conducting comparative trials. Consequently, there is no strong body of material that provides evidence of any specific effect.

Since biodynamic farming is a form of organic farming, it can be generally assumed to share its characteristics, including "less stressed soils and thus diverse and highly interrelated soil communities".

A 2009/2011 review found that biodynamically cultivated fields:

- had lower absolute yields than conventional farms, but achieved better efficiency of production relative to the amount of energy used;

- had greater earthworm populations and biomass than conventional farms.

Both factors were similar to the result in organically cultivated fields.

Reception

In a 2002 newspaper editorial, Peter Treue, agricultural researcher at the University of Kiel, characterized biodynamics as pseudoscience and argued that similar or equal results can be obtained using standard organic farming principles. He wrote that some biodynamic preparations more resemble alchemy or magic akin to geomancy.

In a 1994 analysis, Holger Kirchmann, a soil researcher with the Swedish University of Agricultural Sciences, concluded that Steiner's instructions were occult and dogmatic, and cannot contribute to the development of alternative or sustainable agriculture. According to Kirchmann, many of Steiner's statements are not provable because scientifically clear hypotheses cannot be made from his descriptions. Kirchmann asserted that when methods of biodynamic agriculture were tested scientifically, the results were unconvincing. Further, in a 2004 overview of biodynamic agriculture, Linda Chalker-Scott, a researcher at Washington State University, characterized biodynamics as pseudoscience, writing that Steiner did not use scientific methods to formulate his theory of biodynamics, and that the later addition of valid organic farming techniques has "muddled the discussion" of Steiner's original idea.

Based on the scant scientific testing of biodynamics, Chalker-Scott concluded "no evidence exists" that homeopathic preparations improve the soil.

In Michael Shermer's *The Skeptic Encyclopedia of Pseudoscience*, Dan Dugan says that the way biodynamic preparations are supposed to be implemented are formulated solely on the basis of Steiner's "own insight". Skeptic Brian Dunning writes "the best way to think of 'biodynamic agriculture' would be as a magic spell cast over an entire farm. Biodynamics sees an entire farm as a single organism, with something that they call a life force."

Florian Leiber, Nikolai Fuchs and Hartmut Spieß, researchers at the Goetheanum, have defended the principles of biodynamics and suggested that critiques of biodynamic agriculture which deny it scientific credibility are "not in keeping with the facts...as they take no notice of large areas of biodynamic management and research." Biodynamic farmers are "charged with developing a continuous dialogue between biodynamic science and the natural sciences *sensu stricto*," despite important differences in paradigms, world views, and value systems.

Philosopher of science Michael Ruse has written that followers of biodynamic agriculture rather enjoy the scientific marginalisation that comes from its pseudoscientific basis, revelling both in its esoteric aspects and the impression that they were in the vanguard of the wider anti-science sentiment that has grown in opposition to modern methods such as genetic modification.

Cash Crop

A cotton ball. Cotton is a significant cash crop. According to the National Cotton Council of America, in 2014, China was the world's largest cotton-producing country with an estimated 100,991,000 480-pound bales. India was ranked second at 42,185,000 480-pound bales.

A cash crop is an agricultural crop which is grown for sale to return a profit. It is typically purchased by parties separate from a farm. The term is used to differentiate marketed crops from subsistence crops, which are those fed to the producer's own livestock or grown as food for the producer's family. In earlier times cash crops were usually only a small (but vital) part of a farm's total yield, while today, especially in developed countries, almost all crops are mainly grown for revenue. In the least developed countries, cash crops are usually crops which attract demand in more developed nations, and hence have some export value.

Yerba mate (left, a key ingredient in the beverage known as mate), roasted by the fire, coffee beans (middle) and tea (right) are all used for caffeinated infusions and have cash crop histories.

Prices for major cash crops are set in commodity markets with global scope, with some local variation (termed as "basis") based on freight costs and local supply and demand balance. A consequence of this is that a nation, region, or individual producer relying on such a crop may suffer low prices should a bumper crop elsewhere lead to excess supply on the global markets. This system has been criticized by traditional farmers. Coffee is an example of a product that has been susceptible to significant commodity futures price variations.

Globalization

Issues involving subsidies and trade barriers on such crops have become controversial in discussions of globalization. Many developing countries take the position that the current international trade system is unfair because it has caused tariffs to be lowered in industrial goods while allowing for low tariffs and agricultural subsidies for agricultural goods. This makes it difficult for a developing nation to export its goods overseas, and forces developing nations to compete with imported goods which are exported from developed nations at artificially low prices. The practice of exporting at artificially low prices is known as dumping, and is illegal in most nations. Controversy over this issue led to the collapse of the Cancún trade talks in 2003, when the Group of 22 refused to consider agenda items proposed by the European Union unless the issue of agricultural subsidies was addressed.

Per Climate Zones

Arctic

The Arctic climate is generally not conducive for the cultivation of cash crops. However, one potential cash crop for the Arctic is *Rhodiola rosea*, a hardy plant used as a medicinal herb that grows in the Arctic. There is currently consumer demand for the plant, but the available supply is less than the demand (as of 2011).

Temperate

Cash crops grown in regions with a temperate climate include many cereals (wheat, rye, corn, barley, oats), oil-yielding crops (e.g. rapeseed, mustard seeds), vegetables (e.g. potatoes), tree fruit or top fruit (e.g. apples, cherries) and soft fruit (e.g. strawberries, raspberries).

A tea plantation in the Cameron Highlands in Malaysia

Subtropical

In regions with a subtropical climate, oil-yielding crops (e.g. soybeans) and some vegetables and herbs are the predominant cash crops.

Tropical

In regions with a tropical climate, coffee, cocoa, sugar cane, bananas, oranges, cotton and jute (a soft, shiny vegetable fiber that can be spun into coarse, strong threads), are common cash crops. The oil palm is a tropical palm tree, and the fruit from it is used to make palm oil.

By Continent and Country

Africa

Jatropha curcas is a cash crop used to produce biofuel.

Around 60 percent of African workers are employed in the agricultural sector, with about three-fifths of African farmers being subsistence farmers. For example, in Burkina Faso 85% of its residents (over two million people) are reliant upon cotton production for income, and over half of the country's population lives in poverty. Larger farms tend to grow cash crops such as coffee, tea, cotton, cocoa, fruit and rubber. These farms, typically operated by large corporations, cover tens of square kilometres and employ large numbers of laborers. Subsistence farms provide a source of food and a relatively small income for families, but generally fail to produce enough to make re-investment possible.

The situation in which African nations export crops while a significant amount of people on the continent struggle with hunger has been blamed on developed countries, including the United States, Japan and the European Union. These countries protect their own agricultural sectors, through high import tariffs and offer subsidies to their farmers, which some have contended is leading to the overproduction of commodities such as cotton, grain and milk. The result of this is that the global price of such products is continually reduced until Africans are unable to compete in world markets, except in cash crops that do not grow easily in temperate climates.

Africa has realized significant growth in biofuel plantations, many of which are on lands which were purchased by British companies. *Jatropha curcas* is a cash crop grown for biofuel production in Africa. Some have criticized the practice of raising non-food plants for export while Africa has problems with hunger and food shortages, and some studies have correlated the proliferation of land acquisitions, often for use to grow non-food cash crops with increasing hunger rates in Africa.

Australia

Australia produces significant amounts of lentils. It was estimated in 2010 that Australia would produce approximately 143,000 tons of lentils. Most of Australia's lentil harvest is exported to the Indian subcontinent and the Middle East.

United States

Oranges are a significant U.S. cash crop

Cash cropping in the United States rose to prominence after the baby boomer generation and the end of World War II. It was seen as a way to feed the large population boom and continues to be the main factor in having an affordable food supply in the United States. According to the 1997 U.S. Census of Agriculture, 90% of the farms in the United States are still owned by families, with an additional 6% owned by a partnership. Cash crop farmers have utilized precision agricultural technologies combined with time-tested practices to produce affordable food. Based upon United States Department of Agriculture (USDA) statistics for 2010, states with the highest fruit production quantities are California, Florida and Washington.

Various potato cultivars

Sliced sugarcane, a significant cash crop in Hawaii

Vietnam

Coconut is a cash crop of Vietnam.

Global Cash Crops

Coconut palms are cultivated in more than 80 countries of the world, with a total production of 61 million tonnes per year. The oil and milk derived from it are commonly used in cooking and frying; coconut oil is also widely used in soaps and cosmetics.

Sustainability of Cash Crops

Approximately 70% of the world's food is produced by 500 million smallholder farmers. For their livelihood they depend on the production of cash crops, basic commodities that are hard to differentiate in the market. The great majority (80%) of the world's farms measure 2 hectares or less. These smallholder farmers are mainly found in developing countries and are often unorganized, illiterate or enjoyed only basic education. Smallholder farmers have little bargaining power and incomes are low, leading to a situation in which they cannot invest much in upscaling their businesses. In general, farmers lack access to agricultural inputs and finance, and do not have enough knowledge on good agricultural and business practices. These high level problems are in many cases threatening the future of agricultural sectors and theories start evolving on how to secure a sustainable future for agriculture. Sustainable market transformations are initiated in which industry leaders work together in a pre-competitive environment to change market conditions. Sustainable intensification focuses on facilitating entrepreneurial farmers. To stimulate farm investment projects on access to finance for agriculture are also popping up. One example is the SCOPE methodology, an assessment tool that measures the management maturity and professionalism of producer organizations as to give financing organizations better insights in the risks involved in financing. Currently agricultural finance is always considered risky and avoided by financial institutions.

Black Market Cash Crops

Coca, opium poppies and cannabis are significant black market cash crops, the prevalence of which varies. In the United States, cannabis is considered by some to be the most valuable cash crop. In

2006, it was reported in a study by Jon Gettman, a marijuana policy researcher, that in contrast to government figures for legal crops such as corn and wheat and using the study's projections for U.S. cannabis production at that time, cannabis was cited as "the top cash crop in 12 states and among the top three cash crops in 30." The study also estimated cannabis production at the time (in 2006) to be valued at $35.8 billion USD, which exceeded the combined value of corn at $23.3 billion and wheat at $7.5 billion.

In the U.S., cannabis has been termed as a cash crop.

Permaculture

Permaculture is a system of agricultural and social design principles centered on simulating or directly utilizing the patterns and features observed in natural ecosystems. The term *permaculture* (as a systematic method) was first coined by David Holmgren, then a graduate student, and his professor, Bill Mollison, in 1978. The word *permaculture* originally referred to "permanent agriculture", but was expanded to stand also for "permanent culture", as it was understood that social aspects were integral to a truly sustainable system as inspired by Masanobu Fukuoka's natural farming philosophy.

It has many branches that include but are not limited to ecological design, ecological engineering, environmental design, construction and integrated water resources management that develops sustainable architecture, regenerative and self-maintained habitat and agricultural systems modeled from natural ecosystems.

Mollison has said: "Permaculture is a philosophy of working with, rather than against nature; of protracted and thoughtful observation rather than protracted and thoughtless labor; and of looking at plants and animals in all their functions, rather than treating any area as a single product system."

History

In 1929, Joseph Russell Smith took up an antecedent term as the subtitle for *Tree Crops: A Permanent Agriculture*, a book in which he summed up his long experience experimenting with fruits and nuts as crops for human food and animal feed. Smith saw the world as an inter-related whole and suggested mixed systems of trees and crops underneath. This book inspired many individuals intent on making agriculture more sustainable, such as Toyohiko Kagawa who pioneered forest farming in Japan in the 1930s.

The definition of permanent agriculture as that which can be sustained indefinitely was supported by Australian P. A. Yeomans in his 1964 book *Water for Every Farm*. Yeomans introduced an observation-based approach to land use in Australia in the 1940s, and the keyline design as a way of managing the supply and distribution of water in the 1950s.

Stewart Brand's works were an early influence noted by Holmgren. Other early influences include Ruth Stout and Esther Deans, who pioneered no-dig gardening, and Masanobu Fukuoka who, in the late 1930s in Japan, began advocating no-till orchards, gardens and natural farming.

Core tenets and Principles of Design

The three core tenets of permaculture are:

- Care for the earth: Provision for all life systems to continue and multiply. This is the first principle, because without a healthy earth, humans cannot flourish.

- Care for the people: Provision for people to access those resources necessary for their existence.

- Return of surplus: Reinvesting surpluses back into the system to provide for the first two ethics. This includes returning waste back into the system to recycle into usefulness. The third ethic is sometimes referred to as Fair Share to reflect that each of us should take no more than what we need before we reinvest the surplus.

Permaculture design emphasizes patterns of landscape, function, and species assemblies. It determines where these elements should be placed so they can provide maximum benefit to the local environment. The central concept of permaculture is maximizing useful connections between components and synergy of the final design. The focus of permaculture, therefore, is not on each separate element, but rather on the relationships created among elements by the way they are placed together; the whole becoming greater than the sum of its parts. Permaculture design therefore seeks to minimize waste, human labor, and energy input by building systems with maximal benefits between design elements to achieve a high level of synergy. Permaculture designs evolve over time by taking into account these relationships and elements and can become extremely complex systems that produce a high density of food and materials with minimal input.

The design principles which are the conceptual foundation of permaculture were derived from the science of systems ecology and study of pre-industrial examples of sustainable land use. Permaculture draws from several disciplines including organic farming, agroforestry, integrated farming, sustainable development, and applied ecology. Permaculture has been applied most commonly to the design of housing and landscaping, integrating techniques such as agroforestry, natural building, and rainwater harvesting within the context of permaculture design principles and theory.

Theory

Twelve Design Principles

Twelve Permaculture design principles articulated by David Holmgren in his *Permaculture: Principles and Pathways Beyond Sustainability*:

1. *Observe and interact*: By taking time to engage with nature we can design solutions that suit our particular situation.

2. *Catch and store energy*: By developing systems that collect resources at peak abundance, we can use them in times of need.

3. *Obtain a yield*: Ensure that you are getting truly useful rewards as part of the work that you are doing.

4. *Apply self-regulation and accept feedback*: We need to discourage inappropriate activity to ensure that systems can continue to function well.

5. *Use and value renewable resources and services*: Make the best use of nature's abundance to reduce our consumptive behavior and dependence on non-renewable resources.

6. *Produce no waste*: By valuing and making use of all the resources that are available to us, nothing goes to waste.

7. *Design from patterns to details*: By stepping back, we can observe patterns in nature and society. These can form the backbone of our designs, with the details filled in as we go.

8. *Integrate rather than segregate*: By putting the right things in the right place, relationships develop between those things and they work together to support each other.

9. *Use small and slow solutions*: Small and slow systems are easier to maintain than big ones, making better use of local resources and producing more sustainable outcomes.

10. *Use and value diversity*: Diversity reduces vulnerability to a variety of threats and takes advantage of the unique nature of the environment in which it resides.

11. *Use edges and value the marginal*: The interface between things is where the most interesting events take place. These are often the most valuable, diverse and productive elements in the system.

12. *Creatively use and respond to change*: We can have a positive impact on inevitable change by carefully observing, and then intervening at the right time.

Layers

Suburban permaculture garden in Sheffield, UK with different layers of vegetation

Layers are one of the tools used to design functional ecosystems that are both sustainable and of direct benefit to humans. A mature ecosystem has a huge number of relationships between its component parts: trees, understory, ground cover, soil, fungi, insects, and animals. Because plants grow to different heights, a diverse community of life is able to grow in a relatively small space, as the vegetation occupies different layers. There are generally seven recognized layers in a food forest, although some practitioners also include fungi as an eighth layer.

1. The canopy: the tallest trees in the system. Large trees dominate but typically do not saturate the area, i.e. there exist patches barren of trees.

2. Understory layer: trees that revel in the dappled light under the canopy.

3. Shrub layer: a diverse layer of woody perennials of limited height. includes most berry bushes.

4. Herbaceous layer: Plants in this layer die back to the ground every winter (if winters are cold enough, that is). They do not produce woody stems as the Shrub layer does. Many culinary and medicinal herbs are in this layer. A large variety of beneficial plants fall into this layer. May be annuals, biennials or perennials.

5. Soil surface/Groundcover: There is some overlap with the Herbaceous layer and the Groundcover layer; however plants in this layer grow much closer to the ground, grow densely to fill bare patches of soil, and often can tolerate some foot traffic. Cover crops retain soil and lessen erosion, along with green manures that add nutrients and organic matter to the soil, especially nitrogen.

6. Rhizosphere: Root layers within the soil. The major components of this layer are the soil and the organisms that live within it such as plant roots (including root crops such as potatoes and other edible tubers), fungi, insects, nematodes, worms, etc.

7. Vertical layer: climbers or vines, such as runner beans and lima beans (vine varieties).

Guilds

There are many forms of guilds, including guilds of plants with similar functions (that could inter-change within an ecosystem), but the most common perception is that of a mutual support guild. Such a guild is a group of species where each provides a unique set of diverse functions that work in conjunction, or harmony. Mutual support guilds are groups of plants, animals, insects, etc. that work well together. Some plants may be grown for food production, some have tap roots that draw nutrients up from deep in the soil, some are nitrogen-fixing legumes, some attract beneficial insects, and others repel harmful insects. When grouped together in a mutually beneficial arrangement, these plants form a guild.

Edge effect

The edge effect in ecology is the effect of the juxtaposition or placing side by side of contrasting environments on an ecosystem. Permaculturists argue that, where vastly differing systems meet, there is an intense area of productivity and useful connections. An example of this is the coast; where the land and the sea meet there is a particularly rich area that meets a disproportionate per-centage of human and animal needs. So this idea is played out in permacultural designs by using spirals in the herb garden or creating ponds that have wavy undulating shorelines rather than a simple circle or oval (thereby increasing the amount of edge for a given area).

Zones

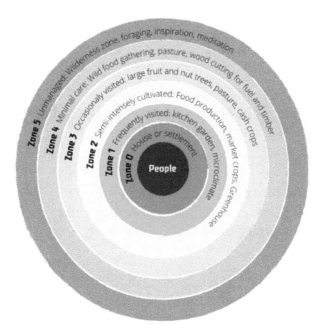

Permaculture Zones 0-5.

Zones are a way of intelligently organizing design elements in a human environment on the basis of the frequency of human use and plant or animal needs. Frequently manipulated or harvested elements of the design are located close to the house in zones 1 and 2. Less frequently used or manipulated elements, and elements that benefit from isolation (such as wild species) are farther away. Zones are about positioning things appropriately, and are numbered from 0 to 5.

Zone 0

The house, or home center. Here permaculture principles would be applied in terms of aiming to reduce energy and water needs, harnessing natural resources such as sunlight, and generally creating a harmonious, sustainable environment in which to live and work. Zone 0 is an informal designation, which is not specifically defined in Bill Mollison's book.

Zone 1

The zone nearest to the house, the location for those elements in the system that require frequent attention, or that need to be visited often, such as salad crops, herb plants, soft fruit like strawberries or raspberries, greenhouse and cold frames, propagation area, worm compost bin for kitchen waste, etc. Raised beds are often used in zone 1 in urban areas.

Zone 2

This area is used for siting perennial plants that require less frequent maintenance, such as occasional weed control or pruning, including currant bushes and orchards, pumpkins, sweet potato, etc. This would also be a good place for beehives, larger scale composting bins, and so on.

Zone 3

The area where main-crops are grown, both for domestic use and for trade purposes. After establishment, care and maintenance required are fairly minimal (provided mulches and similar things are used), such as watering or weed control maybe once a week.

Zone 4

A semi-wild area. This zone is mainly used for forage and collecting wild food as well as production of timber for construction or firewood.

Zone 5

A wilderness area. There is no human intervention in zone 5 apart from the observation of natural ecosystems and cycles. Through this zone we build up a natural reserve of bacteria, moulds and insects that can aid the zones above it.

People and Permaculture

Permaculture uses observation of nature to create regenerative systems, and the place where this has been most visible has been on the landscape. There has been a growing awareness though that firstly, there is the need to pay more attention to the peoplecare ethic, as it is often the dynamics of people that can interfere with projects, and secondly that the principles of permaculture can be used as effectively to create vibrant, healthy and productive people and communities as they have been in landscapes.

Domesticated Animals

Domesticated animals are often incorporated into site design.

Common Practices

Agroforestry

Agroforestry is an integrated approach of using the interactive benefits from combining trees and shrubs with crops and/or livestock. It combines agricultural and forestry technologies to create more diverse, productive, profitable, healthy and sustainable land-use systems. In agroforestry systems, trees or shrubs are intentionally used within agricultural systems, or non-timber forest products are cultured in forest settings.

Forest gardening is a term permaculturalists use to describe systems designed to mimic natural forests. Forest gardens, like other permaculture designs, incorporate processes and relationships that the designers understand to be valuable in natural ecosystems. The terms forest garden and food forest are used interchangeably in the permaculture literature. Numerous permaculturists are proponents of forest gardens, such as Graham Bell, Patrick Whitefield, Dave Jacke, Eric Toensmeier and Geoff Lawton. Bell started building his forest garden in 1991 and wrote the book *The Permaculture Garden* in 1995, Whitefield wrote the book *How to Make a Forest Garden* in 2002, Jacke and Toensmeier co-authored the two volume book set *Edible Forest Gardening* in 2005, and Lawton presented the film *Establishing a Food Forest* in 2008.

Tree Gardens, such as Kandyan tree gardens, in South and Southeast Asia, are often hundreds of years old. Whether they derived initially from experiences of cultivation and forestry, as is the case in agroforestry, or whether they derived from an understanding of forest ecosystems, as is the case for permaculture systems, is not self-evident. Many studies of these systems, especially those that predate the term permaculture, consider these systems to be forms of agroforestry. Permaculturalists who include existing and ancient systems of polycropping with woody species as examples of food forests may obscure the distinction between permaculture and agroforestry.

Food forests and agroforestry are parallel approaches that sometimes lead to similar designs.

Hügelkultur

Hügelkultur is the practice of burying large volumes of wood to increase soil water retention. The porous structure of wood acts as a sponge when decomposing underground. During the rainy season, masses of buried wood can absorb enough water to sustain crops through the dry season. This technique has been used by permaculturalists Sepp Holzer, Toby Hemenway, Paul Wheaton, and Masanobu Fukuoka.

Natural Building

A natural building involves a range of building systems and materials that place major emphasis on sustainability. Ways of achieving sustainability through natural building focus on durability and the use of minimally processed, plentiful or renewable resources, as well as those that, while recycled or salvaged, produce healthy living environments and maintain indoor air quality.

The basis of natural building is the need to lessen the environmental impact of buildings and other supporting systems, without sacrificing comfort, health or aesthetics. To be more sustainable, natural building uses primarily abundantly available, renewable, reused or recycled materials. In addition to relying on natural building materials, the emphasis on the architectural design is heightened. The orientation of a building, the utilization of local climate and site conditions, the emphasis on natural ventilation through design, fundamentally lessen operational costs and positively impact the environment. Building compactly and minimizing the ecological footprint is common, as are on-site handling of energy acquisition, on-site water capture, alternate sewage treatment and water reuse.

Rainwater Harvesting

Rainwater harvesting is the accumulating and storing of rainwater for reuse before it reaches the aquifer. It has been used to provide drinking water, water for livestock, water for irrigation, as well as other typical uses. Rainwater collected from the roofs of houses and local institutions can make an important contribution to the availability of drinking water. It can supplement the subsoil water level and increase urban greenery. Water collected from the ground, sometimes from areas which are especially prepared for this purpose, is called stormwater harvesting.

Greywater is wastewater generated from domestic activities such as laundry, dishwashing, and bathing, which can be recycled on-site for uses such as landscape irrigation and constructed wetlands. Greywater is largely sterile, but not potable (drinkable). Greywater differs from water from the toilets which is designated sewage or blackwater, to indicate it contains human waste. Blackwater is septic or otherwise toxic and cannot easily be reused. There are, however, continuing efforts to make use of blackwater or human waste. The most notable is for composting through a process known as humanure; a combination of the words human and manure. Additionally, the methane in humanure can be collected and used similar to natural gas as a fuel, such as for heating or cooking, and is commonly referred to as biogas. Biogas can be harvested from the human waste and the remainder still used as humanure. Some of the simplest forms of humanure use include a composting toilet or an outhouse or dry bog surrounded by trees that are heavy feeders which can be coppiced for wood fuel. This process eliminates the use of a standard toilet with plumbing.

Sheet Mulching

In agriculture and gardening, mulch is a protective cover placed over the soil. Any material or combination can be used as mulch, such as stones, leaves, cardboard, wood chips, gravel, etc., though in permaculture mulches of organic material are the most common because they perform more functions. These include: absorbing rainfall, reducing evaporation, providing nutrients, increasing organic matter in the soil, feeding and creating habitat for soil organisms, suppressing weed growth and seed germination, moderating diurnal temperature swings, protecting against frost, and reducing erosion. Sheet mulching is an agricultural no-dig gardening technique that attempts to mimic natural processes occurring within forests. Sheet mulching mimics the leaf cover that is found on forest floors. When deployed properly and in combination with other Permacultural principles, it can generate healthy, productive and low maintenance ecosystems.

Sheet mulch serves as a "nutrient bank," storing the nutrients contained in organic matter and slowly making these nutrients available to plants as the organic matter slowly and naturally breaks down. It also improves the soil by attracting and feeding earthworms, slaters and many other soil micro-organisms, as well as adding humus. Earthworms "till" the soil, and their worm castings are among the best fertilizers and soil conditioners. Sheet mulching can be used to reduce or eliminate undesirable plants by starving them of light, and can be more advantageous than using herbicide or other methods of control.

Intensive Rotational Grazing

Grazing has long been blamed for much of the destruction we see in the environment. However, it has been shown that when grazing is modeled after nature, the opposite effect can be seen. Also known as cell grazing, managed intensive rotational grazing (MIRG) is a system of grazing in which ruminant and non-ruminant herds and/or flocks are regularly and systematically moved to fresh pasture, range, or forest with the intent to maximize the quality and quantity of forage growth. This disturbance is then followed by a period of rest which allows new growth. MIRG can be used with cattle, sheep, goats, pigs, chickens, rabbits, geese, turkeys, ducks and other animals depending on the natural ecological community that is being mimicked. Sepp Holzer and Joel Salatin have shown how the disturbance caused by the animals can be the spark needed to start ecological succession or prepare ground for planting. Allan Savory's holistic management technique has been likened to "a permaculture approach to rangeland management". One variation on MIRG that is gaining rapid popularity is called eco-grazing. Often used to either control invasives or re-establish native species, in eco-grazing the primary purpose of the animals is to benefit the environment and the animals can be, but are not necessarily, used for meat, milk or fiber.

Keyline Design

Keyline design is a technique for maximizing beneficial use of water resources of a piece of land developed in Australia by farmer and engineer P. A. Yeomans. The *Keyline* refers to a specific topographic feature linked to water flow which is used in designing the drainage system of the site.

Fruit Tree Management

> The no-pruning option is usually ignored by fruit experts, though often practised by default in people's back gardens! But it has its advantages. Obviously it reduces work, and more surprisingly it can lead to higher overall yields.
> — *Whitefield, Patrick, How to make a forest garden, p. 16*

Masanobu Fukuoka, as part of early experiments on his family farm in Japan, experimented with no-pruning methods, noting that he ended up killing many fruit trees by simply letting them go, which made them become convoluted and tangled, and thus unhealthy.[page needed] Then he realised this is the difference between natural-form fruit trees and the process of change of tree form that results from abandoning previously-pruned unnatural fruit trees.[page needed] He concluded that the trees should be raised all their lives without pruning, so they form healthy and efficient branch patterns that follow their natural inclination. This is part of his implementation of the Tao-philosophy of Wú wéi translated in part as no-action (against nature), and he described it as no unnecessary pruning, nature farming or "do-nothing" farming, of fruit trees, distinct from non-intervention or literal no-pruning. He ultimately achieved yields comparable to or exceeding standard/intensive practices of using pruning and chemical fertilisation.[page needed]

Another proponent of the no, or limited, pruning method is Sepp Holzer who used the method in connection with Hügelkultur berms. He has successfully grown several varieties of fruiting trees at altitudes (approximately 9,000 feet (2,700 m)) far above their normal altitude, temperature, and snow load ranges. He notes that the Hügelkultur berms kept and/or generated enough heat to allow the roots to survive during alpine winter conditions. The point of having unpruned branches, he notes, was that the longer (more naturally formed) branches bend over under the snow load until they touched the ground, thus forming a natural arch against snow loads that would break a shorter, pruned, branch.

Compost Management

Compost Basket

Compost Basket is a way to the permanent management of compost materials. The idea of Compost Basket comes from Gyulai Iván. He invented and used it in the Gömörszőlős Educational Center. The inner circle is 1 meter deep. This is the place where compost material are put in. Under this is the outer circle, 40 cm deep. Here the nutrients come down.

Mollison and Holmgren

Bill Mollison in January 2008.

In the mid-1970s, Bill Mollison and David Holmgren started developing ideas about stable agricultural systems on the southern Australian island state of Tasmania. This was a result of the danger of the rapidly growing use of industrial-agricultural methods. In their view, highly dependent on non renewable resources, these methods were additionally poisoning land and water, reducing biodiversity, and removing billions of tons of topsoil from previously fertile landscapes. A design approach called *permaculture* was their response and was first made public with the publication of their book *Permaculture One* in 1978.

By the early 1980s, the concept had broadened from agricultural systems design towards sustainable human habitats. After *Permaculture One*, Mollison further refined and developed the ideas by designing hundreds of permaculture sites and writing more detailed books, notably *Permaculture: A Designers Manual*. Mollison lectured in over 80 countries and taught his two-week Permaculture Design Course (PDC) to many hundreds of students.Mollison "encouraged graduates to become teachers themselves and set up their own institutes and demonstration sites. This multiplier effect was critical to permaculture's rapid expansion."

In 1991, a four-part television documentary by ABC productions called "The Global Gardener" showed permaculture applied to a range of worldwide situations, bringing the concept to a much broader public. In 2012, the UMass Permaculture Initiative won the White House "Champions of Change" sustainability contest, which declared that "they demonstrate how permaculture can feed a growing population in an environmentally sustainable and socially responsible manner".

In 1997, Holmgren explained that the primary agenda of the permaculture movement is to assist people to become more self-reliant through the design and development of productive and sustainable gardens and farms.

In 2014, Holmgren endorsed and helped launch a new Australian permaculture magazine, Pip Magazine.

Notable Permaculturists

Joseph Russell Smith took up an antecedent term as the subtitle for *Tree Crops: A Permanent Agriculture*, a book in which he summed up his long experience experimenting with fruits and nuts as crops for human food and animal feed. By that year (1929), Smith saw the world as an interrelated whole and suggested mixed systems of trees and crops underneath. This book inspired many individuals intent on making permaculture a valid means of sustainable food production. Bill Mollison and David Holmgren developed it further, and permaculturists were trained under the umbrella of Bill Mollison's train the trainer system.

Geoff Lawton, Toby Hemenway and P. A. Yeomans - creator of the keyline design each have more than 20 years experience teaching and promoting permaculture as a sustainable way of growing food. Simon Fjell was a Founding Director of the Permaculture Institute in late 1979, over 40 years experience, having first met Mollison in 1976. He has since worked in every continent.

The permaculture movement also spread throughout Asia and Central America, with Hong Kong-based Asian Institute of Sustainable Architecture (AISA), Rony Lec leading the foundation of the Mesoamerican Permaculture Institute (IMAP) in Guatemala and Juan Rojas co-founding the Permaculture Institute of El Salvador.

Trademark and Copyright Issues

There has been contention over who, if anyone, controls legal rights to the word *permaculture*: is it trademarked or copyrighted? and if so, who holds the legal rights to the use of the word? For a long time Bill Mollison claimed to have copyrighted the word, and his books said on the copyright page, "The contents of this book and the word PERMACULTURE are copyright." These statements were largely accepted at face-value within the permaculture community. However, copyright law does not protect names, ideas, concepts, systems, or methods of doing something; it only protects the expression or the description of an idea, not the idea itself. Eventually Mollison acknowledged that he was mistaken and that no copyright protection existed for the word *permaculture*.

In 2000, Mollison's US based Permaculture Institute sought a service mark (a form of trademark) for the word *permaculture* when used in educational services such as conducting classes, seminars, or workshops. The service mark would have allowed Mollison and his two Permaculture Institutes (one in the US and one in Australia) to set enforceable guidelines regarding how permaculture could be taught and who could teach it, particularly with relation to the PDC, despite the fact that he had instituted a system of certification of teachers to teach the PDC in 1993. The service mark failed and was abandoned in 2001. Also in 2001 Mollison applied for trademarks in Australia for the terms "Permaculture Design Course" and "Permaculture Design". These applications were both withdrawn in 2003. In 2009 he sought a trademark for "Permaculture: A Designers' Manual" and "Introduction to Permaculture", the names of two of his books. These applications were withdrawn in 2011. There has never been a trademark for the word *permaculture* in Australia.

Criticisms

General Criticisms

In 2011, Owen Hablutzel argued that "permaculture has yet to gain a large amount of specific mainstream scientific acceptance," and that "the sensitiveness to being perceived and accepted on scientific terms is motivated in part by a desire for permaculture to expand and become increasingly relevant." Bec-Hellouin permaculture farm engaged in a research program in partnership with INRA and AgroParisTech to collect scientific data.

In his books *Sustainable Freshwater Aquaculture* and *Farming in Ponds and Dams*, Nick Romanowski expresses the view that the presentation of aquaculture in Bill Mollison's books is unrealistic and misleading.

Agroforestry

Greg Williams argues that forests cannot be more productive than farmland because the net productivity of forests decline as they mature due to ecological succession. Proponents of permaculture respond that this is true only if one compares data between woodland forest and climax vegetation, but not when comparing farmland vegetation with woodland forest. For example, ecological succession generally results in a forest's productivity rising after its establishment only until it reaches the *woodland state* (67% tree cover), before declining until *full maturity*.

Dryland Farming

Dryland farming in the Granada region in Spain

Dryland farming and **dry farming** are agricultural techniques for non-irrigated cultivation of crops. Dryland farming is associated with drylands; dry farming is often associated with areas characterized by a cool wet season followed by a warm dry season.

Dryland Farming Locations

Fields in the Palouse, Washington State

Dryland farming is used in the Great Plains, the Palouse plateau of Eastern Washington, and other arid regions of North America such as in the Southwestern United States and Mexico, the Middle East and in other grain growing regions such as the steppes of Eurasia and Argentina. Dryland farming was introduced to southern Russia and Ukraine by Slavic Mennonites under the influence of Johann Cornies, making the region the breadbasket of Europe. In Australia, it is widely practiced in all states but the Northern Territory.

Dry Farming Locations

Dry farming may be practiced in areas that have significant annual rainfall during a wet season, often in the winter. Crops are cultivated during the subsequent dry season, using practices that make use of the stored moisture in the soil. California, Colorado and Oregon, in the United States, are two states where dry farming is practiced for a variety of crops.

Dryland Farming Crops

Dryland farmed crops may include winter wheat, corn, beans, Sunflowers or even watermelon. Successful dryland farming is possible with as little as 9 inches (230 mm) of precipitation a year; higher rainfall increases the variety of crops. Native American tribes in the arid Southwest survived for hundreds of years on dryland farming in areas with less than 10 inches (250 mm) of rain. The choice of crop is influenced by the timing of the predominant rainfall in relation to the seasons. For example, winter wheat is more suited to regions with higher winter rainfall while areas with summer wet seasons may be more suited to summer growing crops such as sorghum, sunflowers or cotton.

Dry Farmed Crops

Dry farmed crops may include grapes, tomatoes, pumpkins, beans, and other summer crops. These crops grow using the winter water stored in the soil, rather than depending on rainfall during the growing season.

Dryland Farming Process

Dryland farming has evolved as a set of techniques and management practices used by farmers to continually adapt to the presence or lack of moisture in a given crop cycle. In marginal regions, a farmer should be financially able to survive occasional crop failures, perhaps for several years in succession. Survival as a dryland farmer requires careful husbandry of the moisture available for the crop and aggressive management of expenses to minimize losses in poor years. Dryland farming involves the constant assessing of the amount of moisture present or lacking for any given crop cycle and planning accordingly. Dryland farmers know that to be financially successful they have to be aggressive during the good years in order to offset the dry years.

Dryland farming is dependent on natural rainfall, which can leave the ground vulnerable to dust storms, particularly if poor farming techniques are used or if the storms strike at a particularly vulnerable time. The fact that a fallow period must be included in the crop rotation means that fields cannot always be protected by a cover crop, which might otherwise offer protection against erosion.

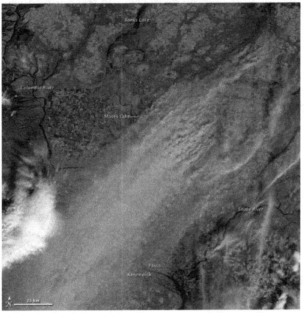

Dryland farming caused a large dust storm in parts of Eastern Washington on October 4, 2009.
Courtesy: NASA/GSFC, MODIS Rapid Response

Dry Farming Process

Dry farming depends on making the best use of the "bank" of soil moisture that was created by winter rainfall. Some dry farming practices include:

- Wider than normal spacing, to provide a larger bank of moisture for each plant.

- Controlled Traffic

- No-till/zero-till or minimum till

- Strict weed control, to ensure that weeds do not consume soil moisture needed by the cultivated plants.

- Cultivation of soil to produce a "dust mulch", thought to prevent the loss of water through capillary action. This practice is controversial, and is not universally advocated.

- Selection of crops and cultivars suited for dry farming practices.

Key Elements of Dryland Farming

Capturing and conservation of moisture – In regions such as Eastern Washington, the average annual precipitation available to a dryland farm may be as little as 8.5 inches (220 mm). Consequently, moisture must be captured until the crop can utilize it. Techniques include summer fallow rotation (in which one crop is grown on two seasons' precipitation, leaving standing stubble and crop residue to trap snow), and preventing runoff by terracing fields.

"Terracing" is also practiced by farmers on a smaller scale by laying out the direction of furrows to slow water runoff downhill, usually by plowing along either contours or keylines. Moisture can be conserved by eliminating weeds and leaving crop residue to shade the soil.

Effective use of available moisture – Once moisture is available for the crop to use, it must be used as effectively as possible. Seed planting depth and timing are carefully considered to place the seed at a depth at which sufficient moisture exists, or where it will exist when seasonal precipitation falls. Farmers tend to use crop varieties which are drought and heat-stress tolerant, (even lower-yielding varieties). Thus the likelihood of a successful crop is hedged if seasonal precipitation fails.

Soil conservation – The nature of dryland farming makes it particularly susceptible to erosion, especially wind erosion. Some techniques for conserving soil moisture (such as frequent tillage to kill weeds) are at odds with techniques for conserving topsoil. Since healthy topsoil is critical to sustainable dryland agriculture, its preservation is generally considered the most important long-term goal of a dryland farming operation. Erosion control techniques such as windbreaks, reduced tillage or no-till, spreading straw (or other mulch on particularly susceptible ground), and strip farming are used to minimize topsoil loss.

Control of input costs – Dryland farming is practiced in regions inherently marginal for non-irrigated agriculture. Because of this, there is an increased risk of crop failure and poor yields which may occur in a dry year (regardless of money or effort expended). Dryland farmers must evaluate the potential yield of a crop constantly throughout the growing season and be prepared to decrease inputs to the crop such as fertilizer and weed control if it appears that it is likely to have a poor yield due to insufficient moisture. Conversely, in years when moisture is abundant, farmers may increase their input efforts and budget to maximize yields and to offset poor harvests.

Arid-zone Agriculture

An example of a dryland farming paddock

As an area of research and development, arid-zone agriculture, or desert agriculture, includes studies of how to increase the agricultural productivity of lands dominated by lack of freshwater, an abundance of heat and sunlight, and usually one or more of extreme winter cold, short rainy season, saline soil or water, strong dry winds, poor soil structure, over-grazing, limited technological development, poverty, political instability.

The two basic approaches are

- view the given environmental and socioeconomic characteristics as negative obstacles to be overcome

- view as many as possible of them as positive resources to be used

Desert Farming

Desert farming generally relies on irrigation, as it is the easiest way to make a desert bloom. In California, the Imperial Valley is a good example of what can be done. Australia and the Horn of Africa are also places with interesting desert agriculture.

Planted forest near Tel Aviv

One problem associated with raising traditional plants in a desert is depletion of the ground water. Drip irrigation is one way to reduce the overall water demand. Another is to grow crops that are acclimated to the desert, such as jojoba, date palms, and citrus.

Native Americans of the Sonoran Desert have long practiced (and continue to practice) desert agriculture *without* irrigation. This is highly dependent upon both winter snow and rain and summer monsoonal weather patterns which move moist tropical air from the Gulf of Mexico into northern Mexico and the southwestern U.S. states of Arizona, Utah, New Mexico, and Colorado. This moist air, combined with the intense solar heating of the ground, can cause the development of substantial thunderstorms that can deluge some portions of the ground with great amounts of water over a short time. While the surface quickly becomes dry, and there is much runoff of the water into normally dry streams and riverbeds, a substantial portion is absorbed into the upper soil levels. To take advantage of this condition for agriculture it is essential that crops be started early in the season, where they utilize residual winter water from Pacific Ocean storms originating in the Gulf of Alaska and other Winter Pacific storms from the tropics (the "Pineapple Express"). The early sprouts from planted seeds (mostly beans, squash, strawberries, and maize) are protected by starting them in small dug holes, where they are both closer to the winter water remaining in the soil and protected from the early spring frosts. Despite its success over a period of (likely) up to 14,000 years, there have been great difficulties with this form of agriculture since a drought beginning in 2002, with global warming suspected in changing weather patterns, and problems with ground water depletion due to extraction for modern conventional irrigated agriculture, metropolitan lawns, industrial purposes, and coal slurry pipelines (the latter now stopped through action by Navajo tribal authority).

The Native Americans practicing this agriculture included the ancient and no longer present Anasazi, the long-present Hopi, the Tewa, Zuni, and many other regional tribes, including the relatively recently arriving (about 1000 to 1400 CE) Navajo. These various tribes were characterized generally by the Spanish occupiers of the region as Sinagua Indians, *sinagua* meaning "without water", although this term is not applied to the modern Native Americans of the region.

Owing to the great dependence upon weather, an element considered to be beyond human control, substantial religious beliefs, rites, and prayer evolved around the growing of crops, and in particular the growing of the four principal corn types of the region, characterized by their colors: red, yellow, blue, and white. The presence of corn as a spiritual symbol can often be seen in the hands of the "Yeh" spirit figures represented in Navajo rugs, in the rituals associated with the "Corn Maiden" and other kachinas of the Hopi, and in various fetish objects of tribes of the region.

American Indians in the Sonoran Desert and elsewhere relied both on irrigation and "Ak-Chin" farming—a type of farming that depended on "washes" (the seasonal flood plains by winter snows and summer rains). The Ak-Chin people employed this natural form of irrigation by planting downslope from a wash, allowing floodwaters to slide over their crops.

In the Salt River Valley, now characterized by Maricopa County, Arizona, a vast canal system that was created and maintined from about 600 AD to 1450 AD. Several hundred miles of canals fed crops of the area surrounding Phoenix, Tempe, Chandler and Mesa, Arizona. The ancient canals served as a model for modern irrigation engineers, with the earliest "modern" historic canals being formed largely by cleaning out the Hohokam canals or being laid out over the top of ancient canals. The ancient ruins and canals of the Hohokam Indians were a source of pride to the early settlers who envisioned their new agricultural society rising as the mythical phoenix bird from the ashes of Hohokam society, hence the name Phoenix, Arizona. The canal system is especially impressive because it was built without the use of metal implements or the wheel. It took remarkable knowledge of geography and hydrology for ancient engineers to lay out the canals, but it also took remarkable socio-political organization to plan workforce deployment, including meeting the physical needs of laborers and their families as well as maintaining and administering the water resources.

Middle Eastern countries are also trying to develop sustainable desert farming.

References

- Stinner, D.H (2007). "The Science of Organic Farming". In William Lockeretz. Organic Farming: An International History. Oxfordshire, UK & Cambridge, Massachusetts: CAB International (CABI). ISBN 978-1-84593-289-3. Retrieved 30 April 2013.

- Vogt G (2007). Lockeretz W, ed. Chapter 1: The Origins of Organic Farming. Organic Farming: An International History. CABI Publishing. pp. 9–30. ISBN 9780851998336.

- Szykitka, Walter (2004). The Big Book of Self-Reliant Living: Advice and Information on Just About Everything You Need to Know to Live on Planet Earth. Globe-Pequot. p. 343. ISBN 978-1-59228-043-8.

- Pamela Ronald; Raoul Admachak (April 2008). "Tomorrow's Table: Organic Farming, Genetics and the Future of Food". Oxford University Press. ISBN 0195301757.

- Halberg, Niels (2006). Global development of organic agriculture: challenges and prospects. CABI. p. 297. ISBN 978-1-84593-078-3.

- Blair, Robert. (2012). Organic Production and Food Quality: A Down to Earth Analysis. Wiley-Blackwell, Oxford, UK. ISBN 978-0-8138-1217-5

- Lejano RP, Ingram M, Ingram HM (2013). "Chapter 6: Narratives of Nature and Science in Alternative Farming Networks". Power of Narrative in Environmental Networks. MIT Press. p. 155. ISBN 9780262519571.

- Vogt G (2007). Lockeretz W, ed. Chapter 1: The Origins of Organic Farming. Organic Farming: An International History. CABI Publishing. pp. 9–30. ISBN 9780851998336.

- Ikerd, John (2010). "Sustainability, Rural". In Leslie A. Duram. Encyclopedia of Organic, Sustainable, and Local Food. ABC-CLIO. pp. 347–349. ISBN 0313359636.

- Abbott, L. K.; Murphy, Daniel V. (2007). Soil Biological Fertility: A Key to Sustainable Land Use in Agriculture. Springer. p. 233. ISBN 140206618X.

- K. Padmavathy; G. Poyyamoli (2011). Lichtfouse, Eric, ed. Genetics, biofuels and local farming system. Berlin: Springer. p. 387. ISBN 978-94-007-1520-2.

- Desai, B K (2007). Sustainable agriculture: a vision for the future. New Delhi: B T Pujari/New India Pub. Agency. pp. 228–9. ISBN 978-81-89422-63-9.

- Dugan D (2002). Shermer M, ed. Anthroposophy and Anthroposophical Medicine. The Skeptic encyclopedia of pseudoscience. 2. ABC-CLIO. p. 32. ISBN 1-57607-653-9.

- Hillstrom, Kevin; Hillstrom, Laurie Collier (2003). Australia, Oceania, and Antarctica: A Continental Overview of Environmental Issues. ABC-CLIO, Inc. pp. 73–119. Retrieved April 9, 2012. ISBN 1576076954

- Holmgren, David (2002). Permaculture: Principles & Pathways Beyond Sustainability. Holmgren Design Services. p. 1. ISBN 0-646-41844-0.

- Nick Romanowski (2007). Sustainable Freshwater Aquaculture: The Complete Guide from Backyard to Investor. UNSW Press. p. 130. ISBN 978-0-86840-835-4.

- Smith, C. Henry (1981). Smith's Story of the Mennonites. Revised and expanded by Cornelius Krahn. Newton, Kansas: Faith and Life Press. pp. 263–265. ISBN 0-87303-069-9.

- Malcolm, Bill; Sale, Peter"; Egan, Adrian (1996). Agriculture in Australia - An Introduction. Australia: Oxford University Press. ISBN 0-19-553695-9.

Soil Fertility: Innovative Practices

Intensive farming practices can sap the essential nutrients that are naturally found in the soil. These nutrients are essential for the growth and developments of plants. In this chapter, techniques such as hydroponics that deliver energy to plants and nutrition restoration are discussed.

Soil Fertility

Soil fertility refers to the ability of a soil to sustain plant growth, i.e. to provide plant habitat and result in sustained and consistent yields of high quality. A fertile soil has the following properties:

- The ability to supply essential plant nutrients and soil water in adequate amounts and proportions for plant growth and reproduction; and

- The absence of toxic substances which may inhibit plant growth.

The following properties contribute to soil fertility in most situations:

- Sufficient soil depth for adequate root growth and water retention;

- Good internal drainage, allowing sufficient aeration for optimal root growth (although some plants, such as rice, tolerate waterlogging);

- Topsoil with sufficient soil organic matter for healthy soil structure and soil moisture retention;

- Soil pH in the range 5.5 to 7.0 (suitable for most plants but some prefer or tolerate more acid or alkaline conditions);

- Adequate concentrations of essential plant nutrients in plant-available forms;

- Presence of a range of microorganisms that support plant growth.

In lands used for agriculture and other human activities, maintenance of soil fertility typically requires the use of soil conservation practices. This is because soil erosion and other forms of soil degradation generally result in a decline in quality with respect to one or more of the aspects indicated above.

Soil scientists use the capital letters O, A, B, C, and E to identify the master horizons, and lowercase letters for distinctions of these horizons. Most soils have three major horizons—the surface horizon (A), the subsoil (B), and the substratum (C). Some soils have an organic horizon (O) on the surface, but this horizon can also be buried. The master horizon, E, is used for subsurface horizons that have a significant loss of minerals (eluviation). Hard bedrock, which is not soil, uses the letter R.

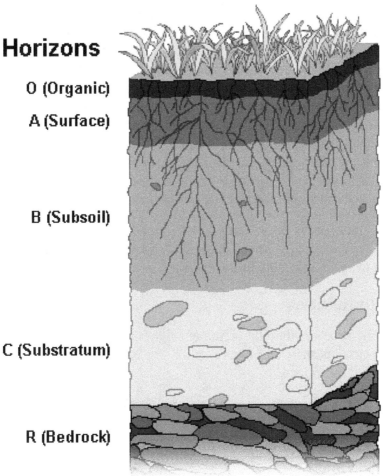

Horizons

O (Organic)

A (Surface)

B (Subsoil)

C (Substratum)

R (Bedrock)

Soil Fertilization

Bioavailable phosphorus is the element in soil that is most often lacking. Nitrogen and potassium are also needed in substantial amounts. For this reason these three elements are always identified on a commercial fertilizer analysis. For example, a 10-10-15 fertilizer has 10 percent nitrogen, 10 percent (P_2O_5) available phosphorus and 15 percent (K_2O) water-soluble potassium. Sulfur is the fourth element that may be identified in a commercial analysis—e.g. 21-0-0-24 which would contain 21% nitrogen and 24% sulfate.

Inorganic fertilizers are generally less expensive and have higher concentrations of nutrients than organic fertilizers. Also, since nitrogen, phosphorus and potassium generally must be in the inorganic forms to be taken up by plants, inorganic fertilizers are generally immediately bioavailable to plants without modification. However, some have criticized the use of inorganic fertilizers, claiming that the water-soluble nitrogen doesn't provide for the long-term needs of the plant and creates water pollution. Slow-release fertilizers may reduce leaching loss of nutrients and may make the nutrients that they provide available over a longer period of time.

Soil fertility is a complex process that involves the constant cycling of nutrients between organic and inorganic forms. As plant material and animal wastes are decomposed by micro-organisms, they release inorganic nutrients to the soil solution, a process referred to as mineralization. Those nutrients may then undergo further transformations which may be aided or enabled by soil micro-organisms. Like plants, many micro-organisms require or preferentially use inorganic forms of nitrogen, phosphorus or potassium and will compete with plants for these nutrients, tying up the nutrients in microbial biomass, a process often called immobilization. The balance between immobilization and mineralization processes depends on the balance and availability of major nutrients and organic carbon to soil microorganisms. Natural processes such as lightning strikes may fix atmospheric nitrogen by converting it to (NO_2). Denitrification may occur under anaerobic conditions (flooding) in the presence of denitrifying bacteria. The cations, primarily phosphate and potash, as well as many micronutrients are held in relatively strong bonds with the negatively charged portions of the soil in a process known as cation-exchange capacity.

In 2008 the cost of phosphorus as fertilizer more than doubled, while the price of rock phosphate as base commodity rose eight-fold. Recently the term peak phosphorus has been coined, due to the limited occurrence of rock phosphate in the world.

Light and CO_2 limitations

Photosynthesis is the process whereby plants use light energy to drive chemical reactions which convert CO_2 into sugars. As such, all plants require access to both light and carbon dioxide to produce energy, grow and reproduce.

While typically limited by nitrogen, phosphorus and potassium, low levels of carbon dioxide can also act as a limiting factor on plant growth. Peer-reviewed and published scientific studies have shown that increasing CO_2 is highly effective at promoting plant growth up to levels over 300 ppm. Further increases in CO_2 can, to a very small degree, continue to increase net photosynthetic output.

Since higher levels of CO_2 have only a minimal impact on photosynthetic output at present levels (presently around 400 ppm and increasing), we should not consider plant growth to be limited by carbon dioxide. Other biochemical limitations, such as soil organic content, nitrogen in the soil, phosphorus and potassium, are far more often in short supply. As such, neither commercial nor scientific communities look to air fertilization as an effective or economic method of increasing production in agriculture or natural ecosystems. Furthermore, since microbial decomposition occurs faster under warmer temperatures, higher levels of CO_2 (which is one of the causes of unusually fast climate change) should be expected to increase the rate at which nutrients are leached out of soils and may have a negative impact on soil fertility.

Soil Depletion

Soil depletion occurs when the components which contribute to fertility are removed and not replaced, and the conditions which support soil's fertility are not maintained. This leads to poor crop yields. In agriculture, depletion can be due to excessively intense cultivation and inadequate soil management.

Soil fertility can be severely challenged when land use changes rapidly. For example, in Colonial New England, colonists made a number of decisions that depleted the soils, including: allowing herd animals to wander freely, not replenishing soils with manure, and a sequence of events that led to erosion. William Cronon wrote that "...the long-term effect was to put those soils in jeopardy. The removal of the forest, the increase in destructive floods, the soil compaction and close-cropping wrought by grazing animals, plowing--all served to increase erosion."

One of the most widespread occurrences of soil depletion as of 2008 is in tropical zones where nutrient content of soils is low. The combined effects of growing population densities, large-scale industrial logging, slash-and-burn agriculture and ranching, and other factors, have in some places depleted soils through rapid and almost total nutrient removal.

Topsoil depletion occurs when the nutrient-rich organic topsoil, which takes hundreds to thousands of years to build up under natural conditions, is eroded or depleted of its original organic material. Historically, many past civilizations' collapses can be attributed to the depletion of the topsoil. Since the beginning of agricultural production in the Great Plains of North America in the 1880s, about one-half of its topsoil has disappeared.

Depletion may occur through a variety of other effects, including overtillage (which damages soil structure), underuse of nutrient inputs which leads to mining of the soil nutrient bank, and salinization of soil.

Irrigation Water Effects

The quality of irrigation water is very important to maintain soil fertility and tilth, and for using more soil depth by the plants. When soil is irrigated with high alkaline water, unwanted sodium salts build up in the soil which would make soil draining capacity very poor. So plant roots can not penetrate deep in to the soil for optimum growth in Alkali soils. When soil is irrigated with low pH / acidic water, the useful salts (Ca, Mg, K, P, S, etc.) are removed by draining water from the acidic soil and in addition unwanted aluminium and manganese salts to the plants are dissolved from the soil impeding plant growth. When soil is irrigated with high salinity water or sufficient water is not draining out from the irrigated soil, the soil would convert in to saline soil or lose its fertility. Saline water enhance the turgor pressure or osmotic pressure requirement which impedes the off take of water and nutrients by the plant roots.

Top soil loss takes place in alkali soils due to erosion by rain water surface flows or drainage as they form colloids (fine mud) in contact with water. Plants absorb water-soluble inorganic salts only from the soil for their growth. Soil as such does not lose fertility just by growing crops but it lose its fertility due to accumulation of unwanted and depletion of wanted inorganic salts from the soil by improper irrigation and acid rain water (quantity and quality of water). The fertility of many soils which are not suitable for plant growth can be enhanced many times gradually by providing adequate irrigation water of suitable quality and good drainage from the soil.

Global Distribution

Global distribution of soil types of the USDA soil taxonomy system. Mollisols, shown here in dark green, are a good (though not the only) indicator of high soil fertility. They coincide to a large extent with the world's major grain producing areas like the North American Prairie States, the Pampa and Gran Chaco of South America and the Ukraine-to-Central Asia Black Earth belt.

Global Soil Regions

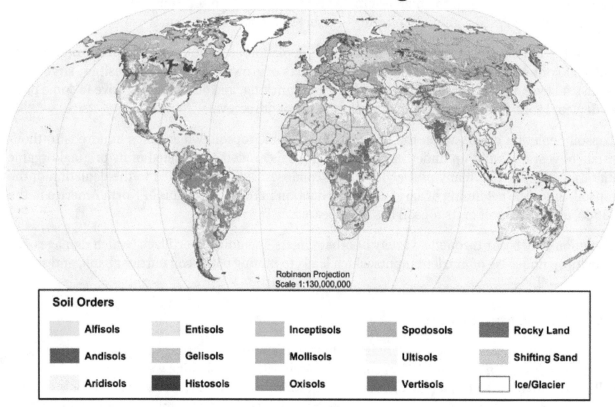

USDA

US Department of Agriculture   Soil Survey Division
Natural Resources   World Soil Resources
Conservation Service   soils.usda.gov/use/worldsoils

November 2005

Plant Nutrition

Plant nutrition is the study of the chemical elements and compounds necessary for plant growth, plant metabolism and their external supply. In 1972, E. Epstein defined two criteria for an element to be essential for plant growth:

1. in its absence the plant is unable to complete a normal life cycle.

2. or that the element is part of some essential plant constituent or metabolite.

This is in accordance with Justus von Liebig's law of the minimum. The essential plant nutrients include carbon and oxygen which are absorbed from the air, whereas other nutrients including hydrogen are typically obtained from the soil (exceptions include some parasitic or carnivorous plants).

Plants must obtain the following mineral nutrients from their growing medium:

- the macronutrients: nitrogen (N), phosphorus (P), potassium (K), calcium (Ca), sulfur (S), magnesium (Mg); and

- the micronutrients (or trace minerals): boron (B), chlorine (Cl), manganese (Mn), iron (Fe), zinc (Zn), copper (Cu), molybdenum (Mo), nickel (Ni).

The macronutrients are consumed in larger quantities and are usually present in plant tissue in concentrations of between 0.2% and 4.0% (on a dry matter weight basis). Micronutrients are present in plant tissue in quantities measured in parts per million, ranging from 0.1 to 200 ppm, or less than 0.02% dry weight.

Farmer spreading decomposing manure to improve soil fertility and plant nutrition

Most soil conditions across the world can provide plants adapted to that climate and soil with sufficient nutrition for a complete life cycle, without the addition of nutrients as fertilizer. However, if the soil is cropped it is necessary to artificially modify soil fertility through the addition of fertilizer to promote vigorous growth and increase or sustain yield. This is done because, even with adequate water and light, nutrient deficiency can limit growth and crop yield.

Processes

Plants take up essential elements from the soil through their roots and from the air (mainly consisting of nitrogen and oxygen) through their leaves. Nutrient uptake in the soil is achieved by cation exchange, wherein root hairs pump hydrogen ions (H^+) into the soil through proton pumps. These hydrogen ions displace cations attached to negatively charged soil particles so that the cations are available for uptake by the root. In the leaves, stomata open to take in carbon dioxide and expel oxygen. The carbon dioxide molecules are used as the carbon source in photosynthesis.

The root, especially the root hair, is the essential organ for the uptake of nutrients. The structure and architecture of the root can alter the rate of nutrient uptake. Nutrient ions are transported to the center of the root, the stele in order for the nutrients to reach the conducting tissues, xylem and phloem. The Casparian strip, a cell wall outside the stele but within the root, prevents passive flow of water and nutrients, helping to regulate the uptake of nutrients and water. Xylem moves water and inorganic molecules within the plant and phloem accounts for organic molecule transportation. Water potential plays a key role in a plant's nutrient uptake. If the water potential is more negative within the plant than the surrounding soils, the nutrients will move from the region of higher solute concentration—in the soil—to the area of lower solute concentration - in the plant.

There are three fundamental ways plants uptake nutrients through the root:

1. Simple diffusion occurs when a nonpolar molecule, such as O_2, CO_2, and NH_3 follows a concentration gradient, moving passively through the cell lipid bilayer membrane without the use of transport proteins.

2. Facilitated diffusion is the rapid movement of solutes or ions following a concentration gradient, facilitated by transport proteins.

3. Active transport is the uptake by cells of ions or molecules against a concentration gradient; this requires an energy source, usually ATP, to power molecular pumps that move the ions or molecules through the membrane.

Nutrients can be moved within plants to where they are most needed. For example, a plant will try to supply more nutrients to its younger leaves than to its older ones. When nutrients are mobile within the plant, symptoms of any deficiency become apparent first on the older leaves. However, not all nutrients are equally mobile. Nitrogen, phosphorus, and potassium are mobile nutrients while the others have varying degrees of mobility. When a less-mobile nutrient is deficient, the younger leaves suffer because the nutrient does not move up to them but stays in the older leaves. This phenomenon is helpful in determining which nutrients a plant may be lacking.

Many plants engage in symbiosis with microorganisms. Two important types of these relationship are

1. with bacteria such as rhizobia, that carry out biological nitrogen fixation, in which atmospheric nitrogen (N_2) is converted into ammonium (NH_4^+); and

2. with mycorrhizal fungi, which through their association with the plant roots help to create a larger effective root surface area. Both of these mutualistic relationships enhance nutrient uptake.

Though nitrogen is plentiful in the Earth's atmosphere, relatively few plants harbour nitrogen-fixing bacteria, so most plants rely on nitrogen compounds present in the soil to support their growth. These can be supplied by mineralization of soil organic matter or added plant residues, nitrogen fixing bacteria, animal waste, through the breaking of triple bonded Nitrogen molecules by lightening strikes or through the application of fertilizers.

Functions of Nutrients

At least 17 elements are known to be essential nutrients for plants. In relatively large amounts, the soil supplies nitrogen, phosphorus, potassium, calcium, magnesium, and sulfur; these are often called the macronutrients. In relatively small amounts, the soil supplies iron, manganese, boron, molybdenum, copper, zinc, chlorine, and cobalt, the so-called micronutrients. Nutrients must be available not only in sufficient amounts but also in appropriate ratios.

Plant nutrition is a difficult subject to understand completely, partially because of the variation between different plants and even between different species or individuals of a given clone. Elements present at low levels may cause deficiency symptoms, and toxicity is possible at levels that are too high. Furthermore, deficiency of one element may present as symptoms of toxicity from another element, and vice versa. An abundance of one nutrient may cause a deficiency of another nutrient. For example, K^+ uptake can be influenced by the amount of NH_4^+ available.

Although nitrogen is plentiful in the Earth's atmosphere, relatively few plants engage in nitrogen fixation (conversion of atmospheric nitrogen to a biologically useful form). Most plants, therefore, require nitrogen compounds to be present in the soil in which they grow.

Carbon and oxygen are absorbed from the air while other nutrients are absorbed from the soil. Green plants obtain their carbohydrate supply from the carbon dioxide in the air by the process of photosynthesis. Each of these nutrients is used in a different place for a different essential function.

Macronutrients (Derived from Air and Water)

Carbon

Carbon forms the backbone of most plant biomolecules, including proteins, starches and cellulose. Carbon is fixed through photosynthesis; this converts carbon dioxide from the air into carbohydrates which are used to store and transport energy within the plant.

Hydrogen

Hydrogen also is necessary for building sugars and building the plant. It is obtained almost entirely from water. Hydrogen ions are imperative for a proton gradient to help drive the electron transport chain in photosynthesis and for respiration.

Oxygen

Oxygen is a component of many organic and inorganic molecules within the plant, and is acquired in many forms. These include: O_2 and CO_2 (mainly from the air via leaves) and H_2O, NO_3^-, $H_2PO_4^-$ and SO_4^{2-} (mainly from the soil water via roots). Plants produce oxygen gas (O_2) along with glucose during photosynthesis but then require O_2 to undergo aerobic cellular respiration and break down this glucose to produce ATP.

Nitrogen

Nitrogen is a major constituent of several of the most important plant substances. For example, nitrogen compounds comprise 40% to 50% of the dry matter of protoplasm, and it is a constituent of amino acids, the building blocks of proteins. It is also an essential constituent of chlorophyll. Nitrogen deficiency most often results in stunted growth, slow growth, and chlorosis. Nitrogen deficient plants will also exhibit a purple appearance on the stems, petioles and underside of leaves from an accumulation of anthocyanin pigments. Most of the nitrogen taken up by plants is from the soil in the forms of NO_3^-, although in acid environments such as boreal forests where nitrification is less likely to occur, ammonium NH_4^+ is more likely to be the dominating source of nitrogen. Amino acids and proteins can only be built from NH_4^+, so NO_3^- must be reduced. In many agricultural settings, nitrogen is the limiting nutrient for rapid growth. Nitrogen is transported via the xylem from the roots to the leaf canopy as nitrate ions, or in an organic form, such as amino acids or amides. Nitrogen can also be transported in the phloem sap as amides, amino acids and ureides; it is therefore mobile within the plant, and the older leaves exhibit chlorosis and necrosis earlier than the younger leaves.

There is an abundant supply of nitrogen in the earth's atmosphere — N_2 gas comprises nearly 79% of air. However, N_2 is unavailable for use by most organisms because there is a triple bond between the two nitrogen atoms, making the molecule almost inert. In order for nitrogen to be used for growth it must be "fixed" (combined) in the form of ammonium (NH_4) or nitrate (NO_3) ions. The weathering of rocks releases these ions so slowly that it has a negligible effect on the availability of fixed nitrogen. Therefore, nitrogen is often the limiting factor for growth and biomass production in all environments where there is a suitable climate and availability of water to support life.

Nitrogen enters the plant largely through the roots. A "pool" of soluble nitrogen accumulates. Its composition within a species varies widely depending on several factors, including day length, time of day, night temperatures, nutrient deficiencies, and nutrient imbalance. Short day length promotes asparagine formation, whereas glutamine is produced under long day regimes. Darkness favors protein breakdown accompanied by high asparagine accumulation. Night temperature modifies the effects due to night length, and soluble nitrogen tends to accumulate owing to retarded synthesis and breakdown of proteins. Low night temperature conserves glutamine; high night temperature increases accumulation of asparagine because of breakdown. Deficiency of K accentuates differences between long- and short-day plants. The pool of soluble nitrogen is much smaller than in well-nourished plants when N and P are deficient since uptake of nitrate and further reduction and conversion of N to organic forms is restricted more than is protein synthesis. Deficiencies of Ca, K, and S affect the conversion of organic N to protein more than uptake and reduction. The size of the pool of soluble N is no guide *per se* to growth rate, but the size of the pool in relation to total N might be a useful ratio in this regard. Nitrogen availability in the rooting medium also affects the size and structure of tracheids formed in the long lateral roots of white spruce (Krasowski and Owens 1999).

Microorganisms have a central role in almost all aspects of nitrogen availability, and therefore for life support on earth. Some bacteria can convert N_2 into ammonia by the process termed *nitrogen fixation*; these bacteria are either free-living or form symbiotic associations with plants or other organisms (e.g., termites, protozoa), while other bacteria bring about transformations of ammonia to nitrate, and of nitrate to N_2 or other nitrogen gases. Many bacteria and fungi degrade organic matter, releasing fixed nitrogen for reuse by other organisms. All these processes contribute to the nitrogen cycle.

Phosphorus

Like nitrogen, phosphorus is involved with many vital plant processes. Within a plant, it is present mainly as a structural component of the nucleic acids, deoxyribonucleic nucleic acid (DNA) and ribose nucleic acid (RNA), and as a constituent of fatty phospholipids, of importance in membrane development and function. It is present in both organic and inorganic forms, both of which are readily translocated within the plant. All energy transfers in the cell are critically dependent on phosphorus. As with all living things, phosphorus is part of the Adenosine triphosphate (ATP), which is of immediate use in all processes that require energy with the cells. Phosphorus can also be used to modify the activity of various enzymes by phosphorylation, and is used for cell signaling. Phosphorus is concentrated at the most actively growing points of a plant and stored within seeds in anticipation of their germination. Phosphorus is most commonly found in the soil in the form of polyprotic phosphoric acid (H_3PO_4), but is taken up most readily in the form of $H_2PO_4^-$. Phosphorus is available to plants in limited quantities in most soils because it is released very slowly from insoluble phosphates and is rapidly fixed once again. Under most environmental conditions it is the element that limits growth because of this constriction and due to its high demand by plants and microorganisms. Plants can increase phosphorus uptake by a mutualism with mycorrhiza. A Phosphorus deficiency in plants is characterized by an intense green coloration or reddening in leaves due to lack of chlorophyll. If the plant is experiencing high phosphorus deficiencies the leaves may become denatured and show signs of death. Occasionally the leaves may appear purple from an accumulation of anthocyanin. Because phosphorus is a mobile nutrient, older leaves will show the first signs of deficiency.

On some soils, the phosphorus nutrition of some conifers, including the spruces, depends on the ability of mycorrhizae to take up, and make soil phosphorus available to the tree, hitherto unobtainable to the non-mycorrhizal root. Seedling white spruce, greenhouse-grown in sand testing negative for phosphorus, were very small and purple for many months until spontaneous mycorrhizal inoculation, the effect of which was manifested by a greening of foliage and the development of vigorous shoot growth.

Phosphorus deficiency can produce symptoms similar to those of nitrogen deficiency, but as noted by Russel: "Phosphate deficiency differs from nitrogen deficiency in being extremely difficult to diagnose, and crops can be suffering from extreme starvation without there being any obvious signs that lack of phosphate is the cause". Russell's observation applies to at least some coniferous seedlings, but Benzian found that although response to phosphorus in very acid forest tree nurseries in England was consistently high, no species (including Sitka spruce) showed any visible symptom of deficiency other than a slight lack of lustre. Phosphorus levels have to be exceedingly low before visible symptoms appear in such seedlings. In sand culture at 0 ppm phosphorus, white spruce seedlings were very small and tinted deep purple; at 0.62 ppm, only the smallest seedlings were deep purple; at 6.2 ppm, the seedlings were of good size and color.

It is useful to apply a high phosphorus content fertilizer, such as bone meal, to perennials to help with successful root formation.

Potassium

Unlike other major elements, potassium does not enter into the composition of any of the important plant constituents involved in metabolism, but it does occur in all parts of plants in substantial amounts. It seems to be of particular importance in leaves and at growing points. Potassium is outstanding among the nutrient elements for its mobility and solubility within plant tissues. Processes involving potassium include the formation of carbohydrates and proteins, the regulation of internal plant moisture, as a catalyst and condensing agent of complex substances, as an accelerator of enzyme action, and as contributor to photosynthesis, especially under low light intensity.

When soil-potassium levels are high, plants take up more potassium than needed for healthy growth. The term *luxury consumption* has been applied to this. When potassium is moderately deficient, the effects first appear in the older tissues, and from there progress towards the growing points. Acute deficiency severely affects growing points, and die-back commonly occurs. Symptoms of potassium deficiency in white spruce include: browning and death of needles (chlorosis); reduced growth in height and diameter; impaired retention of needles; and reduced needle length. A relationship between potassium nutrition and cold resistance has been found in several tree species, including 2 species of spruce.

Potassium regulates the opening and closing of the stomata by a potassium ion pump. Since stomata are important in water regulation, potassium reduces water loss from the leaves and increases drought tolerance. Potassium deficiency may cause necrosis or interveinal chlorosis. K^+ is highly mobile and can aid in balancing the anion charges within the plant. Potassium helps in fruit coloration, shape and also increases its brix. Hence, quality fruits are produced in potassium-rich soils. Potassium serves as an activator of enzymes used in photosynthesis and respiration Potassium is used to build cellulose and aids in photosynthesis by the formation of a chlorophyll precursor. Potassium deficiency may result in higher risk of pathogens, wilting, chlorosis, brown spotting, and higher chances of damage from frost and heat.

Macronutrients (Secondary and Tertiary)

Sulfur

Sulfur is a structural component of some amino acids and vitamins, and is essential in the manufacturing of chloroplasts. Sulfur is also found in the iron-sulfur complexes of the electron transport chains in photosynthesis. It is immobile and deficiency, therefore, affects younger tissues first. Symptoms of deficiency include yellowing of leaves and stunted growth.

Calcium

Calcium regulates transport of other nutrients into the plant and is also involved in the activation of certain plant enzymes. Calcium deficiency results in stunting. This nutrient is involved in photosynthesis and plant structure. Blossom end rot is also a result of inadequate calcium.

Calcium in plants occurs chiefly in the leaves, with lower concentrations in seeds, fruits, and roots. A major function is as a constituent of cell walls. When coupled with certain acidic compounds of the jelly-like pectins of the middle lamella, calcium forms an insoluble salt. It is also intimately involved in meristems, and is particularly important in root development, with roles in cell division, cell elongation, and the detoxification of hydrogen ions. Other functions attributed to calcium are; the neutralization of organic acids; inhibition of some potassium-activated ions; and a role in nitrogen absorption. A notable feature of calcium-deficient plants is a defective root system. Roots are usually affected before above-ground parts.

Magnesium

The outstanding role of magnesium in plant nutrition is as a constituent of the chlorophyll molecule. As a carrier, it is also involved in numerous enzyme reactions as an effective activator, in which it is closely associated with energy-supplying phosphorus compounds. Magnesium is very mobile in plants, and, like potassium, when deficient is translocated from older to younger tissues, so that signs of deficiency appear first on the oldest tissues and then spread progressively to younger tissues.

Micro-Nutrients

Some elements are directly involved in plant metabolism (Arnon and Stout, 1939). However, this principle does not account for the so-called beneficial elements, whose presence, while not required, has clear positive effects on plant growth. Mineral elements that either stimulate growth but are not essential, or that are essential only for certain plant species, or under given conditions, are usually defined as beneficial elements.

Plants are able sufficiently to accumulate most trace elements. Some plants are sensitive indicators of the chemical environment in which they grow (Dunn 1991), and some plants have barrier mechanisms that exclude or limit the uptake of a particular element or ion species, e.g., alder twigs commonly accumulate molybdenum but not arsenic, whereas the reverse is true of spruce bark (Dunn 1991). Otherwise, a plant can integrate the geochemical signature of the soil mass permeated by its root system together with the contained groundwaters. Sampling is facilitated by the tendency of many elements to accumulate in tissues at the plant's extremities.

Iron

Iron is necessary for photosynthesis and is present as an enzyme cofactor in plants. Iron deficiency can result in interveinal chlorosis and necrosis. Iron is not a structural part of chlorophyll but very much essential for its synthesis. Copper deficiency can be responsible for promoting an iron deficiency.

Molybdenum

Molybdenum is a cofactor to enzymes important in building amino acids and is involved in nitrogen metabolism. Molybdenum is part of the nitrate reductase enzyme (needed for the reduction of nitrate) and the nitrogenase enzyme (required for biological nitrogen fixation).

Boron

Boron is found in the highly insoluble mineral, tourmaline. It is absorbed by plants in the form of the anion BO_3^{3-}. It is available to plants in moderately soluble mineral forms of Ca, Mg and Na borates and the highly soluble form of organic compounds. Concentration in soil must, in general, be below 5 ppm in the soil water solution, above that toxicity results. Its availability in soils ranges from 20 to 200 pounds per acre in the first eight inches, of which less than 5% is available. It is available to plants over a range of pH, from 5.0 to 7.5. It is mobile in the soil, hence, it is prone to leaching. Leaching removes substantial amounts of boron in sandy soil, but little in fine silt or clay soil. Boron's fixation to those minerals at high pH can render boron unavailable, while low pH frees the fixed boron, leaving it prone to leaching in wet climates. It precipitates with other minerals in the form of borax in which form it was first used over 400 years ago as a soil supplement. Decomposition of organic material causes boron to be deposited in the topmost soil layer; organic forms of boron are more soluble than their mineral form, hence are more available in the top few inches. When soil dries it can cause a precipitous drop in the availability of boron to plants as the plants cannot draw nutrients from that desiccated layer. Hence, boron deficiency diseases appear in dry weather.

Boron has many functions within a plant: it affects flowering and fruiting, pollen germination, cell division, and active salt absorption. The metabolism of amino acids and proteins, carbohydrates, calcium, and water are strongly affected by boron. Many of those listed functions may be embodied by its function in moving the highly polar sugars through cell membranes by reducing their polarity and hence the energy needed to pass the sugar. If sugar cannot pass to the fastest growing parts rapidly enough, those parts die. Boron is relatively immobile within a plant suggesting that the molecule is fixed to the points in the membrane where they facilitate sugar transport.

Boron is not relocatable in the plant via the phloem. It must be supplied to the growing parts via the xylem. Foliar sprays affect only those parts sprayed, which may be insufficient for the fastest growing parts, and is very temporary.

Boron is essential for the proper forming and strengthening of cell walls. Lack of boron results in short thick cells producing stunted fruiting bodies and roots. Calcium to boron ratio must be maintained in a narrow range for normal plant growth. For alfalfa, that calcium to boron ratio must be from 80:1 to 600:1. Boron deficiency appears at 800:1 and higher. For alfalfa, similar ratios exist for magnesium, copper, nitrogen and potassium. Boron levels within plants differ with plant species and range from 2.3 p.p.m for barley to 94.7 p.p.m for poppy . Lack of boron causes failure of calcium metabolism which produces hollow heart in beets and peanuts.

Inadequate amounts of boron affect many agricultural crops, legume forage crops most strongly. Of the micronutrients, boron deficiencies are second most common after zinc. Deficiencies of boron when soil is cropped are common and require the application of mineral supplement; one ton of alfalfa hay carries with it one ounce of boron, 100 bushels of peaches 4 ounces. Deficiency results in the death of the terminal growing points. Symptoms first appear as stunted growth, then to cellular changes, which leads to physical changes, and finally death of the plant.

Boron supplements derive from dry lake bed deposits such as those in Death Valley, USA, in the form of sodium tetraborate, from which less soluble calcium borate is made. Foliar sprays are used on fruit crop trees in soils of high alkalinity. Boron is often applied to fields as a contaminant in other soil amendments but is not generally adequate to make up the rate of loss by cropping. The rates of application of borate to produce an adequate alfalfa crop range from 15 pounds per acre for a sandy-silt, acidic soil of low organic matter, to 60 pounds per acre for a soil with high organic matter, high cation exchange capacity and high pH.

Boron concentration in soil water solution higher than one ppm is toxic to most plants. Toxic concentrations within plants are 10 to 50 ppm for small grains and 200 ppm in boron-tolerant crops such as sugar beets, rutabaga, cucumbers, and conifers. Toxic soil conditions are generally limited to arid regions or can be caused by underground borax deposits in contact with water or volcanic gases dissolved in percolating water. Application rates should be limited to a few pounds per acre in a test plot to determine if boron is needed generally. Otherwise, testing for boron levels in plant material is required to determine remedies. Excess boron can be removed by irrigation and assisted by application of elemental sulfur to lower the pH and increase boron's solubility. Application of calcium will increase soil alkalinity, causing boron to fix on the mineral soil component and remove some fraction, thereby reducing boron toxicity.

Boron deficiencies must be detected by analysis of plant material to apply a correction before the obvious symptoms appear, after which it is too late to prevent crop loss. Strawberries deficient in boron will produce lumpy fruit; apricots will not blossom or, if they do, will not fruit or will drop their fruit depending on the level of boron deficit. Broadcast of boron supplements is effective and long term; a foliar spray is immediate but must be repeated.

Boron is an essential element for the health of animals which derive their boron from plant material.

Copper

Copper is important for photosynthesis. Symptoms for copper deficiency include chlorosis.It is involved in many enzyme processes; necessary for proper photosynthesis; involved in the manufacture of lignin (cell walls) and involved in grain production. It is also hard to find in some soil conditions.

Manganese

Manganese is necessary for photosynthesis, including the building of chloroplasts. Manganese deficiency may result in coloration abnormalities, such as discolored spots on the foliage.

Sodium

Sodium is involved in the regeneration of phosphoenolpyruvate in CAM and C4 plants. Sodium can potentially replace potassium's regulation of stomatal opening and closing.

Essentiality of sodium:

- Essential for C4 plants rather C3

- Substitution of K by Na: Plants can be classified into four groups:

1. Group A—a high proportion of K can be replaced by Na and stimulate the growth, which cannot be achieved by the application of K

2. Group B—specific growth responses to Na are observed but they are much less distinct

3. Group C—Only minor substitution is possible and Na has no effect

4. Group D—No substitution occurs

- Stimulate the growth—increase leaf area and stomata. Improves the water balance

- Na functions in metabolism

1. C4 metabolism

2. Impair the conversion of pyruvate to phosphoenol-pyruvate

3. Reduce the photosystem II activity and ultrastructural changes in mesophyll chloroplast

- Replacing K functions

1. Internal osmoticum

2. Stomatal function

3. Photosynthesis

4. Counteraction in long distance transport

5. Enzyme activation

- Improves the crop quality e.g. improves the taste of carrots by increasing sucrose

Zinc

Zinc is required in a large number of enzymes and plays an essential role in DNA transcription. A typical symptom of zinc deficiency is the stunted growth of leaves, commonly known as "little leaf" and is caused by the oxidative degradation of the growth hormone auxin.

Nickel

In higher plants, nickel is absorbed by plants in the form of Ni^{2+} ion. Nickel is essential for activation of urease, an enzyme involved with nitrogen metabolism that is required to process urea. Without nickel, toxic levels of urea accumulate, leading to the formation of necrotic lesions. In lower plants, nickel activates several enzymes involved in a variety of processes, and can substitute for zinc and iron as a cofactor in some enzymes.

Chlorine

Chlorine, as compounded chloride, is necessary for osmosis and ionic balance; it also plays a role in photosynthesis.

Cobalt

Cobalt has proven to be beneficial to at least some plants although it does not appear to be essential for most species. It has, however, been shown to be essential for nitrogen fixation by the nitrogen-fixing bacteria associated with legumes and other plants.

Aluminium

- Tea has a high tolerance for aluminum (Al) toxicity and the growth is stimulated by Al application. The possible reason is the prevention of Cu, Mn or P toxicity effects.

- There have been reports that Al may serve as a fungicide against certain types of root rot.

Silicon

Silicon is not considered an essential element for plant growth and development. It is always found in abundance in the environment and hence if needed it is available. It is found in the structures of plants and improves the health of plants.

In plants, silicon has been shown in experiments to strengthen cell walls, improve plant strength, health, and productivity. There have been studies showing evidence of silicon improving drought and frost resistance, decreasing lodging potential and boosting the plant's natural pest and disease fighting systems. Silicon has also been shown to improve plant vigor and physiology by improving root mass and density, and increasing above ground plant biomass and crop yields. Silicon is currently under consideration by the Association of American Plant Food Control Officials (AAPFCO) for elevation to the status of a "plant beneficial substance".

Higher plants differ characteristically in their capacity to take up silicon. Depending on their SiO_2 content they can be divided into three major groups:

- Wetland graminae-wetland rice, horsetail (10–15%)

- Dryland graminae-sugar cane, most of the cereal species and few dicotyledons species (1–3%)

- Most of dicotyledons especially legumes (<0.5%)

- The long distance transport of Si in plants is confined to the xylem. Its distribution within the shoot organ is therefore determined by transpiration rate in the organs

- The epidermal cell walls are impregnated with a film layer of silicon and effective barrier against water loss, cuticular transpiration rate in the organs.

Vanadium

Vanadium may be required by some plants, but at very low concentrations. It may also be substituting for molybdenum.

Selenium

Selenium is probably not essential for flowering plants, but it can be beneficial; it can stimulate plant growth, improve tolerance of oxidative stress, and increase resistance to pathogens and herbivory.

Selenium is, however, an essential mineral element for animal (including human) nutrition and selenium deficiencies are known to occur when food or animal feed is grown on selenium-deficient soils. The use of inorganic selenium fertilizers can increase selenium concentrations in edible crops and animal diets thereby improving animal health.

Nutrient Deficiency

The effect of a nutrient deficiency can vary from a subtle depression of growth rate to obvious stunting, deformity, discoloration, distress, and even death. Visual symptoms distinctive enough to be useful in identifying a deficiency are rare. Most deficiencies are multiple and moderate. However, while a deficiency is seldom that of a single nutrient, nitrogen is commonly the nutrient in shortest supply.

Chlorosis of foliage is not always due to mineral nutrient deficiency. Solarization can produce superficially similar effects, though mineral deficiency tends to cause premature defoliation, whereas solarization does not, nor does solarization depress nitrogen concentration.

Nutrient Status of Plants

Nutrient status (mineral nutrient and trace element composition, also called ionome and nutrient profile) of plants are commonly portrayed by tissue elementary analysis. Interpretation of the results of such studies, however, has been controversial. During the last decades the nearly two-century-old "law of minimum" or "Liebig's law" (that states that plant growth is controlled not by the total amount of resources available, but by the scarcest resource) has been replaced by several mathematical approaches that use different models in order to take the interactions between the individual nutrients into account. The latest developments in this field are based on the fact that the nutrient elements (and compounds) do not act independently from each other; Baxter, 2015, because there may be direct chemical interactions between them or they may influence each other's uptake, translocation, and biological action via a number of mechanisms as exemplified for the case of ammonia.

Plant Nutrition in Agricultural Systems

Hydroponics

Hydroponics is a method for growing plants in a water-nutrient solution without the use of nutrient-rich soil. It allows researchers and home gardeners to grow their plants in a controlled environment. The most common solution is the Hoagland solution, developed by D. R. Hoagland in 1933. The solution consists of all the essential nutrients in the correct proportions necessary for most plant growth. An aerator is used to prevent an anoxic event or hypoxia. Hypoxia can affect nutrient uptake of a plant because, without oxygen present, respiration becomes inhibited within the root cells. The nutrient film technique is a hydroponic technique in which the roots are not fully submerged. This allows for adequate aeration of the roots, while a "film" thin layer of nutrient-rich water is pumped through the system to provide nutrients and water to the plant.

Hydroponics

NASA researcher checking hydroponic onions with Bibb lettuce to his left and radishes to the right

Hydroponics is a subset of hydroculture and is a method of growing plants using mineral nutrient solutions, in water, without soil. Terrestrial plants may be grown with their roots in the mineral solution only, or in an inert medium, such as perlite or gravel. The nutrients in hydroponics can be from fish waste, normal nutrients, or duck manure.

History

The earliest published work on growing terrestrial plants without soil was the 1627 book *Sylva Sylvarum* by Francis Bacon, printed a year after his death. Water culture became a popular research technique after that. In 1699, John Woodward published his water culture experiments with spearmint. He found that plants in less-pure water sources grew better than plants in distilled water. By 1842, a list of nine elements believed to be essential for plant growth had been compiled, and the discoveries of German botanists Julius von Sachs and Wilhelm Knop, in the years 1859–1875, resulted in a development of the technique of soilless cultivation. Growth of terrestrial plants without soil in mineral nutrient solutions was called solution culture. It quickly became a standard research and teaching technique and is still widely used. Solution culture is now considered a type of hydroponics where there is no inert medium.

In 1929, William Frederick Gericke of the University of California at Berkeley began publicly promoting that solution culture be used for agricultural crop production. He first termed it aquaculture but later found that aquaculture was already applied to culture of aquatic organisms. Gericke created a sensation by growing tomato vines twenty-five feet high in his back yard in mineral nutrient solutions rather than soil. He introduced the term hydroponics, water culture, in 1937, proposed to him by W. A. Setchell, a phycologist with an extensive education in the classics. Hydroponics is derived from neologism, constructed in analogy to geoponica, that which concerns agriculture, replacing, earth, with water.

Reports of Gericke's work and his claims that hydroponics would revolutionize plant agriculture prompted a huge number of requests for further information. Gericke had been denied use of the University's greenhouses for his experiments due to the administration's skepticism, and when the University tried to compel him to release his preliminary nutrient recipes developed at home he requested greenhouse space and time to improve them using appropriate research facilities. While he was eventually provided greenhouse space, the University assigned Hoagland and Arnon to re-develop Gericke's formula and show it held no benefit over soil grown plant yields, a view held by Hoagland. In 1940, Gericke published the book, *Complete Guide to Soil less Gardening,* after leaving his academic position in a climate that was politically unfavorable.

Two other plant nutritionists at the University of California were asked to research Gericke's claims. Dennis R. Hoagland and Daniel I. Arnon wrote a classic 1938 agricultural bulletin, *The Water Culture Method for Growing Plants Without Soil,*. Hoagland and Arnon claimed that hydroponic crop yields were no better than crop yields with good-quality soils. Crop yields were ultimately limited by factors other than mineral nutrients, especially light. This research, however, overlooked the fact that hydroponics has other advantages including the fact that the roots of the plant have constant access to oxygen and that the plants have access to as much or as little water as they need. This is important as one of the most common errors when growing is over- and under- watering; and hydroponics prevents this from occurring as large amounts of water can be made available to the plant and any water not used, drained away, recirculated, or actively aerated, eliminating anoxic conditions, which drown root systems in soil. In soil, a grower needs to be very experienced to know exactly how much water to feed the plant. Too much and the plant will be unable to access oxygen; too little and the plant will lose the ability to transport nutrients, which are typically moved into the roots while in solution. These two researchers developed several formulas for mineral nutrient solutions, known as Hoagland solution. Modified Hoagland solutions are still in use.

One of the earliest successes of hydroponics occurred on Wake Island, a rocky atoll in the Pacific Ocean used as a refuelling stop for Pan American Airlines. Hydroponics was used there in the 1930s to grow vegetables for the passengers. Hydroponics was a necessity on Wake Island because there was no soil, and it was prohibitively expensive to airlift in fresh vegetables.

In the 1960s, Allen Cooper of England developed the Nutrient film technique. The Land Pavilion at Walt Disney World's EPCOT Center opened in 1982 and prominently features a variety of hydroponic techniques. In recent decades, NASA has done extensive hydroponic research for its Controlled Ecological Life Support System (CELSS). Hydroponics intended to take place on Mars are using LED lighting to grow in a different color spectrum with much less heat.

Origin

Soilless Culture

Gericke originally defined hydroponics as crop growth in mineral nutrient solutions. Hydroponics is a subset of soilless culture. Many types of soilless culture do not use the mineral nutrient solutions required for hydroponics.

Plants that are not traditionally grown in a climate would be possible to grow using a controlled environment system like hydroponics. NASA has also looked to utilize hydroponics in the space program. Ray Wheeler, a plant physiologist at Kennedy Space Center's Space Life Science Lab, believes that hydroponics will create advances within space travel. He terms this as a bioregenerative life support system.

Techniques

There are two main variations for each medium, sub-irrigation and top irrigation. For all techniques, most hydroponic reservoirs are now built of plastic, but other materials have been used including concrete, glass, metal, vegetable solids, and wood. The containers should exclude light to prevent algae growth in the nutrient solution.

Static Solution Culture

In static solution culture, plants are grown in containers of nutrient solution, such as glass Mason jars (typically, in-home applications), plastic buckets, tubs, or tanks. The solution is usually gently aerated but may be un-aerated. If un-aerated, the solution level is kept low enough that enough roots are above the solution so they get adequate oxygen. A hole is cut in the lid of the reservoir for each plant. There can be one to many plants per reservoir. Reservoir size can be increased as plant– size increases. A home made system can be constructed from plastic food containers or glass canning jars with aeration provided by an aquarium pump, aquarium airline tubing and aquarium valves. Clear containers are covered with aluminium foil, butcher paper, black plastic, or other material to exclude light, thus helping to eliminate the formation of algae. The nutrient solution is changed either on a schedule, such as once per week, or when the concentration drops below a certain level as determined with an electrical conductivity meter. Whenever the solution is depleted below a certain level, either water or fresh nutrient solution is added. A Mariotte's bottle, or a float valve, can be used to automatically maintain the solution level. In raft solution culture, plants are placed in a sheet of buoyant plastic that is floated on the surface of the nutrient solution. That way, the solution level never drops below the roots.

The deep water raft tank at the CDC South Aquaponics greenhouse in Brooks, Alberta.

Continuous-Flow Solution Culture

The nutrient film technique being used to grow various salad greens

In continuous-flow solution culture, the nutrient solution constantly flows past the roots. It is much easier to automate than the static solution culture because sampling and adjustments to the temperature and nutrient concentrations can be made in a large storage tank that has potential to serve thousands of plants. A popular variation is the nutrient film technique or NFT, whereby a very shallow stream of water containing all the dissolved nutrients required for plant growth is re-circulated past the bare roots of plants in a watertight thick root mat, which develops in the bottom of the channel and has an upper surface that, although moist, is in the air. Subsequent to this, an abundant supply of oxygen is provided to the roots of the plants. A properly designed NFT system is based on using the right channel slope, the right flow rate, and the right channel length. The main advantage of the NFT system over other forms of hydroponics is that the plant roots are exposed to adequate supplies of water, oxygen, and nutrients. In all other forms of production, there is a conflict between the supply of these requirements, since excessive or deficient amounts of one results in an imbalance of one or both of the others. NFT, because of its design, provides a system where all three requirements for healthy plant growth can be met at the same time, provided that the simple concept of NFT is always remembered and practised. The result of these advantages is that higher yields of high-quality produce are obtained over an extended period of cropping. A downside of NFT is that it has very little buffering against interruptions in the flow (e.g. power outages). But, overall, it is probably one of the more productive techniques.

The same design characteristics apply to all conventional NFT systems. While slopes along channels of 1:100 have been recommended, in practice it is difficult to build a base for channels that is sufficiently true to enable nutrient films to flow without ponding in locally depressed areas. As a consequence, it is recommended that slopes of 1:30 to 1:40 are used. This allows for minor irregularities in the surface, but, even with these slopes, ponding and water logging may occur. The slope may be provided by the floor, or benches or racks may hold the channels and provide the required slope. Both methods are used and depend on local requirements, often determined by the site and crop requirements.

As a general guide, flow rates for each gully should be 1 liter per minute. At planting, rates may be half this and the upper limit of 2 L/min appears about the maximum. Flow rates beyond these extremes are often associated with nutritional problems. Depressed growth rates of many crops have been observed when channels exceed 12 metres in length. On rapidly growing crops, tests have indicated that, while oxygen levels remain adequate, nitrogen may be depleted over the length of the gully. As a consequence, channel length should not exceed 10–15 metres. In situations where this is not possible, the reductions in growth can be eliminated by placing another nutrient feed halfway along the gully and halving the flow rates through each outlet.

Aeroponics

Aeroponics is a system wherein roots are continuously or discontinuously kept in an environment saturated with fine drops (a mist or aerosol) of nutrient solution. The method requires no substrate and entails growing plants with their roots suspended in a deep air or growth chamber with the roots periodically wetted with a fine mist of atomized nutrients. Excellent aeration is the main advantage of aeroponics.

Aeroponic techniques have proven to be commercially successful for propagation, seed germination, seed potato production, tomato production, leaf crops, and micro-greens. Since inventor Richard Stoner commercialized aeroponic technology in 1983, aeroponics has been implemented as an alternative to water intensive hydroponic systems worldwide. The limitation of hydroponics is the fact that 1 kilogram (2.2 lb) of water can only hold 8 milligrams (0.12 gr) of air, no matter whether aerators are utilized or not.

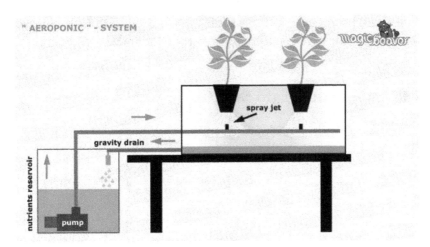

A diagram of the aeroponic technique.

Another distinct advantage of aeroponics over hydroponics is that any species of plants can be grown in a true aeroponic system because the micro environment of an aeroponic can be finely controlled. The limitation of hydroponics is that only certain species of plants can survive for so long in water before they become waterlogged. The advantage of aeroponics is that suspended aeroponic plants receive 100% of the available oxygen and carbon dioxide to the roots zone, stems, and leaves, thus accelerating biomass growth and reducing rooting times. NASA research has shown that aeroponically grown plants have an 80% increase in dry weight biomass (essential minerals) compared to hydroponically grown plants. Aeroponics used 65% less water than hydroponics. NASA also concluded that aeroponically grown plants requires ¼ the nutrient input compared to hydroponics. Unlike hydroponically grown plants, aeroponically grown plants will not suffer transplant shock when transplanted to soil, and offers growers the ability to reduce the spread of disease and pathogens. Aeroponics is also widely used in laboratory studies of plant physiology and plant pathology. Aeroponic techniques have been given special attention from NASA since a mist is easier to handle than a liquid in a zero-gravity environment.

Fogponics

Fogponics is a derivation of aeroponics wherein the nutrient solution is aerosolized by a diaphragm vibrating at ultrasonic frequencies. Solution droplets produced by this method tend to be 5-10 μm in diameter, smaller than those produced by forcing a nutrient solution through pressurized nozzles, as in aeroponics. The smaller size of the droplets allows them to diffuse through the air more easily, and deliver nutrients to the roots without limiting their access to oxygen.

Passive Sub-irrigation

Passive sub-irrigation, also known as passive hydroponics or semi-hydroponics, is a method wherein plants are grown in an inert porous medium that transports water and fertilizer to the roots by capillary action from a separate reservoir as necessary, reducing labor and providing a constant supply of water to the roots. In the simplest method, the pot sits in a shallow solution of fertilizer and water or on a capillary mat saturated with nutrient solution. The various hydroponic media available, such as expanded clay and coconut husk, contain more air space than more traditional potting mixes, delivering increased oxygen to the roots, which is important in epiphytic plants such as orchids and bromeliads, whose roots are exposed to the air in nature. Additional advantages of passive hydroponics are the reduction of root rot and the additional ambient humidity provided through evaporations.

Ebb and Flow or Flood and Drain Sub-Irrigation

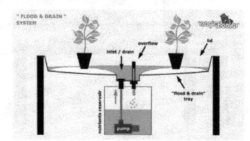

A Ebb and flow or flood and drain hydroponics system.

In its simplest form, there is a tray above a reservoir of nutrient solution. Either the tray is filled with growing medium (clay granules being the most common) and planted directly or pots of medium stand in the tray. At regular intervals, a simple timer causes a pump to fill the upper tray with nutrient solution, after which the solution drains back down into the reservoir. This keeps the medium regularly flushed with nutrients and air. Once the upper tray fills past the drain stop, it begins recirculating the water until the timer turns the pump off, and the water in the upper tray drains back into the reservoirs.

Run to Waste

In a run-to-waste system, nutrient and water solution is periodically applied to the medium surface. The method was invented in Bengal in 1946, for this reason it is sometimes referred to as "The Bengal System".

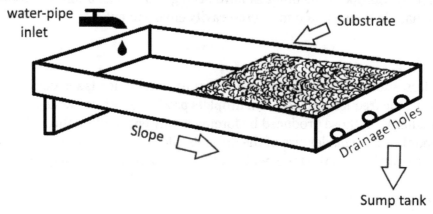

A run-to-waste hydroponics system referred to as "The Bengal System" after the region in northeastern India where it was invented (circa 1946–1948).

This method can be setup in various configurations. In its simplest form, a nutrient-and-water solution is manually applied one or more times per day to a container of inert growing media, such as rockwool, perlite, vermiculite, coco fibre, or sand. In a slightly more complex system, it is automated with a delivery pump, a timer and irrigation tubing to deliver nutrient solution with a delivery frequency that is governed by the key parameters of plant size, plant growing stage, climate, substrate, and substrate conductivity, pH, and water content.

In a commercial setting, watering frequency is multi-factorial and governed by computers or PLCs.

Commercial hydroponics production of large plants like tomatoes, cucumber, and peppers use one form or another of run-to-waste hydroponics.

In environmentally responsible uses, the nutrient rich waste is collected and processed through an on site filtration system to be used many times, making the system very productive.

The majority of bonsai are now grown in soil-free substrates (typically consisting of akadama, grit, diatomaceous earth and other inorganic components) and have their water and nutrients provided in a run-to-waste form.

Deep Water Culture

The Deep water culture technique being used to grow Hungarian wax peppers.

The hydroponic method of plant production by means of suspending the plant roots in a solution of nutrient-rich, oxygenated water. Traditional methods favor the use of plastic buckets and large containers with the plant contained in a net pot suspended from the centre of the lid and the roots suspended in the nutrient solution. The solution is oxygen saturated by an air pump combined with porous stones. With this method, the plants grow much faster because of the high amount of oxygen that the roots receive.

Top-Fed Deep Water Culture

Top-fed deep water culture is a technique involving delivering highly oxygenated nutrient solution direct to the root zone of plants. While deep water culture involves the plant roots hanging down into a reservoir of nutrient solution, in top-fed deep water culture the solution is pumped from the reservoir up to the roots (top feeding). The water is released over the plant's roots and then runs back into the reservoir below in a constantly recirculating system. As with deep water culture, there is an airstone in the reservoir that pumps air into the water via a hose from outside the reservoir. The airstone helps add oxygen to the water. Both the airstone and the water pump run 24 hours a day.

The biggest advantage of top-fed deep water culture over standard deep water culture is increased growth during the first few weeks. With deep water culture, there is a time when the roots have not reached the water yet. With top-fed deep water culture, the roots get easy access to water from the beginning and will grow to the reservoir below much more quickly than with a deep water culture system. Once the roots have reached the reservoir below, there is not a huge advantage with top-fed deep water culture over standard deep water culture. However, due to the quicker growth in the beginning, grow time can be reduced by a few weeks.

Rotary

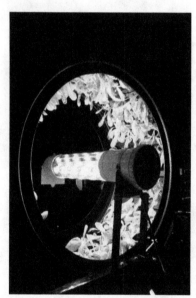

A Rotary hydroponic cultivation demonstration at the Belgian Pavilion Expo in 2015.

A rotary hydroponic garden is a style of commercial hydroponics created within a circular frame which rotates continuously during the entire growth cycle of whatever plant is being grown.

While system specifics vary, systems typically rotate once per hour, giving a plant 24 full turns within the circle each 24-hour period. Within the center of each rotary hydroponic garden is a high intensity grow light, designed to simulate sunlight, often with the assistance of a mechanized timer.

Each day, as the plants rotate, they are periodically watered with a hydroponic growth solution to provide all nutrients necessary for robust growth. Due to the plants continuous fight against gravity, plants typically mature much more quickly than when grown in soil or other traditional hydroponic growing systems. Due to the small foot print a rotary hydroponic system has, it allows for more plant material to be grown per sq foot of floor space than other traditional hydroponic systems.

Substrates

One of the most obvious decisions hydroponic farmers have to make is which medium they should use. Different media are appropriate for different growing techniques.

Expanded Clay Aggregate

Expanded clay pebbles.

Baked clay pellets are suitable for hydroponic systems in which all nutrients are carefully controlled in water solution. The clay pellets are inert, pH neutral and do not contain any nutrient value.

The clay is formed into round pellets and fired in rotary kilns at 1,200°C (2,190°F). This causes the clay to expand, like popcorn, and become porous. It is light in weight, and does not compact over time. The shape of an individual pellet can be irregular or uniform depending on brand and manufacturing process. The manufacturers consider expanded clay to be an ecologically sustainable and re-usable growing medium because of its ability to be cleaned and sterilized, typically by washing in solutions of white vinegar, chlorine bleach, or hydrogen peroxide (H_2O_2), and rinsing completely.

Another view is that clay pebbles are best not re-used even when they are cleaned, due to root growth that may enter the medium. Breaking open a clay pebble after a crop has been grown will reveal this growth.

Growstones

Growstones, made from glass waste, have both more air and water retention space than perlite and peat. This aggregate holds more water than parboiled rice hulls. Growstones by volume consists of .5 to 5% Calcium carbonate. for a standard 5.1 kg bag of Growstones that's 25.8 to 258 grams of calcium carbonate. The remainder is Soda-lime glass.

Coir Peat

Coco peat, also known as coir or coco, is the leftover material after the fibres have been removed from the outermost shell (bolster) of the coconut. Coir is a 100% natural grow and flowering medium. Coconut coir is colonized with trichoderma fungi, which protects roots and stimulates root growth. It is extremely difficult to over-water coir due to its perfect air-to-water ratio; plant roots thrive in this environment. Coir has a high cation exchange, meaning it can store unused minerals to be released to the plant as and when it requires it. Coir is available in many forms; most common is coco peat, which has the appearance and texture of soil but contains no mineral content.

Rice Husks

Rice husks, a hydroponic growing substrate option.

Parboiled rice husks (PBH) are an agricultural byproduct that would otherwise have little use. They decay over time, and allow drainage, and even retain less water than growstones. A study showed that rice husks did not affect the effects of plant growth regulators.

Perlite

Perlite, a hydroponic growing substrate option.

Perlite is a volcanic rock that has been superheated into very lightweight expanded glass pebbles. It is used loose or in plastic sleeves immersed in the water. It is also used in potting soil mixes to decrease soil density. Perlite has similar properties and uses to vermiculite but, in general, holds more air and less water. If not contained, it can float if flood and drain feeding is used. It is a fusion of granite, obsidian, pumice and basalt. This volcanic rock is naturally fused at high temperatures undergoing what is called "Fusionic Metamorphosis".

Vermiculite

Vermiculite close-up.

Like perlite, vermiculite is a mineral that has been superheated until it has expanded into light pebbles. Vermiculite holds more water than perlite and has a natural "wicking" property that can draw water and nutrients in a passive hydroponic system. If too much water and not enough air surrounds the plants roots, it is possible to gradually lower the medium's water-retention capability by mixing in increasing quantities of perlite.

Pumice

A pumice stone.

Like perlite, pumice is a lightweight, mined volcanic rock that finds application in hydroponics.

Sand

Sand is cheap and easily available. However, it is heavy, does not hold water very well, and it must be sterilized between uses.

Gravel

The same type that is used in aquariums, though any small gravel can be used, provided it is washed first. Indeed, plants growing in a typical traditional gravel filter bed, with water circulated using electric powerhead pumps, are in effect being grown using gravel hydroponics. Gravel is inexpensive, easy to keep clean, drains well and will not become waterlogged. However, it is also heavy, and, if the system does not provide continuous water, the plant roots may dry out.

Wood Fibre

Excelsior, or wood wool

Wood fibre, produced from steam friction of wood, is a very efficient organic substrate for hydro-ponics. It has the advantage that it keeps its structure for a very long time. Wood wool (i.e. wood slivers) have been used since the earliest days of the hydroponics research. However, more recent research suggests that wood fibre may have detrimental effects on "plant growth regulators".

Sheep Wool

Wool from shearing sheep is a little-used yet promising renewable growing medium. In a study comparing wool with peat slabs, coconut fibre slabs, perlite and rockwool slabs to grow cucumber plants, sheep wool had a greater air capacity of 70%, which decreased with use to a comparable 43%, and water capacity that increased from 23% to 44% with use. Using sheep wool resulted in the greatest yield out of the tested substrates, while application of a biostimulator consisting of humic acid, lactic acid and Bacillus subtilis improved yields in all substrates.

Rock Wool

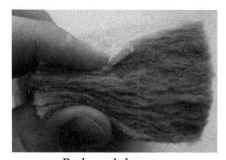

Rock wool close-up.

Rock wool (mineral wool) is the most widely used medium in hydroponics. Rock wool is an in-ert substrate suitable for both run-to-waste and recirculating systems. Rock wool is made from molten rock, basalt or 'slag' that is spun into bundles of single filament fibres, and bonded into a medium capable of capillary action, and is, in effect, protected from most common microbiological degradation. Rock wool has many advantages and some disadvantages. The latter being the possible skin irritancy (mechanical) whilst handling (1:1000). Flushing with cold water usually brings relief. Advantages include its proven efficiency and effectiveness as a commercial hydroponic substrate. Most of the rock wool sold to date is a non-hazardous, non-carcinogenic material, falling under Note Q of the European Union Classification Packaging and Labeling Regulation (CLP).

Brick Shards

Brick shards have similar properties to gravel. They have the added disadvantages of possibly altering the pH and requiring extra cleaning before reuse.

Polystyrene Packing Peanuts

Polystyrene foam peanuts

Polystyrene packing peanuts are inexpensive, readily available, and have excellent drainage. However, they can be too lightweight for some uses. They are used mainly in closed-tube systems. Note that polystyrene peanuts must be used; biodegradable packing peanuts will decompose into a sludge. Plants may absorb styrene and pass it to their consumers; this is a possible health risk.

Nutrient Solutions

Inorganic Hydroponic Solutions

The formulation of hydroponic solutions is an application of plant nutrition, with nutrient deficiency symptoms mirroring those found in traditional soil based agriculture. However, the underlying chemistry of hydroponic solutions can differ from soil chemistry in many significant ways. Important differences include:

- Unlike soil, hydroponic nutrient solutions do not have cation-exchange capacity (CEC) from clay particles or organic matter. The absence of CEC means the pH and nutrient concentrations can change much more rapidly in hydroponic setups than is possible in soil.

- Selective absorption of nutrients by plants often imbalances the amount of counterions in solution. This imbalance can rapidly affect solution pH and the ability of plants to absorb nutrients of similar ionic charge. For instance, nitrate anions are often consumed rapidly by plants to form proteins, leaving an excess of cations in solution. This cation imbalance can lead to deficiency symptoms in other cation based nutrients (e.g. Mg^{2+}) even when an ideal quantity of those nutrients are dissolved in the solution.

- Depending the on pH, and/or the presence of water contaminants, nutrients, such as iron, can precipitate from the solution and become unavailable to plants. Routine adjustments to pH, buffering the solution, and/or the use of chelating agents is often necessary.

As in conventional agriculture, nutrients should be adjusted to satisfy Liebig's law of the minimum for each specific plant variety. Nevertheless, generally acceptable concentrations for nutrient solutions exist, with minimum and maximum concentration ranges for most plants being somewhat similar. Most nutrient solutions are mixed to have concentrations between 1,000 and 2,500 ppm. Acceptable concentrations for the individual nutrient ions, which comprise that total ppm figure, are summarized in the following table. For essential nutrients, concentrations below these ranges often lead to nutrient deficiencies while exceeding these ranges can lead to nutrient toxicity. Optimum nutrition concentrations for plant varieties are found empirically by experience and/or by plant tissue tests.

Element	Role	Ionic form(s)	Low range (ppm)	High range (ppm)	Common Sources	Comment
Nitrogen	Essential macronutrient	NO_3^- and/ or NH_4^+	100	1000	KNO_3, NH_4NO_3, $Ca(NO_3)_2$, HNO_3, $(NH_4)_2SO_4$, and $(NH_4)_2HPO_4$	NH_4^+ interferes with Ca^{2+} uptake and can be toxic to plants if used as a major nitrogen source. A 3:1 ratio of NO_3^- to NH_4^+ is sometimes recommended to balance pH during nitrogen absorption.
Potassium	Essential macronutrient	K^+	100	400	KNO_3, K_2SO_4, KCl, KOH, K_2CO_3, K_2H-PO_4, and K_2SiO_3	High concentrations interfere with the function Fe, Mn, and Zn. Zinc deficiencies often are the most apparent.
Phosphorus	Essential macronutrient	PO_4^{3-}	30	100	K_2HPO_4, KH_2PO_4, $NH_4H_2PO_4$, H_3PO_4, and $Ca(H_2PO_4)_2$	Excess NO_3^- tends to inhibit PO_4^{3-} absorption. The ratio of iron to PO_4^{3-} can affect co-precipitiation reactions.
Calcium	Essential macronutrient	Ca^{2+}	200	500	$Ca(NO_3)_2$, $Ca(H_2PO_4)$, $CaSO_4$, $CaCl_2$	Excess Ca^{2+} inhibits Mg^{2+} uptake.
Magnesium	Essential macronutrient	Mg^{2+}	50	100	$MgSO_4$ and $MgCl_2$	Should not exceed Ca^{2+} concentration due to competitive uptake.
Sulfur	Essential macronutrient	SO_4^{2-}	50	1000	$MgSO_4$, K_2SO_4, $CaSO_4$, H_2SO_4, $(NH_4)_2SO_4$, $ZnSO_4$, $CuSO_4$, $FeSO_4$, and $MnSO_4$	Unlike most nutrients, plants can tolerate a high concentration of the SO_4^{2-}, selectively absorbing the nutrient as needed. Undesirable counterion affects still apply however.

Element	Type	Ion	Min	Max	Sources	Notes
Iron	Essential micronutrient	Fe^{3+} and Fe^{2+}	2	5	FeDTPA, FeEDTA, iron citrate, iron tartrate, $FeCl_3$, and $FeSO_4$	pH values above 6.5 greatly decreases iron solubility. Chelating agents (e.g. DTPA, citric acid, or EDTA) are often added to increase iron solubility over a greater pH range.
Zinc	Essential micronutrient	Zn^{2+}	0.05	1	$ZnSO_4$	Excess zinc is highly toxic to plants but is essential for plants at low concentrations.
Copper	Essential micronutrient	Cu^{2+}	0.01	1	$CuSO_4$	Plant sensitivity to copper is highly variable. 0.1 ppm can cause toxic to some plants while a concentration up to 0.5 ppm for many plants is often considered ideal.
Manganese	Essential micronutrient	Mn^{2+}	0.5	1	$MnSO_4$ and $MnCl_2$	Uptake is enhanced by high PO_4^{3-} concentrations.
Boron	Essential micronutrient	$B(OH)_4^-$	0.3	10	H_3BO_3, and $Na_2B_4O_7$	An essential nutrient, however, some plants are highly sensitive to boron (e.g. toxic affects are apparent in citrus trees at 0.5 ppm).
Molybdenum	Essential micronutrient	MoO_4^-	0.001	0.05	$(NH_4)_6Mo_7O_{24}$ and Na_2MoO_4	A component of the enzyme nitrate reductase and required by rhizobia for nitrogen fixation.
Nickel	Essential micronutrient	Ni^{2+}	0.057	1.5	$NiSO_4$ and $NiCO_3$	Essential to many plants (e.g. legumes and some grain crops). Also used in the enzyme urease.
Chlorine	Variable micronutrient	Cl^-	0	Highly variable	KCl, $CaCl_2$, $MgCl_2$, and $NaCl$	Can interfere with NO_3^- uptake in some plants but can be beneficial in some plants (e.g. in asparagus at 5 ppm). Absent in conifers, ferns, and most bryophytes.
Aluminum	Variable micronutrient	Al^{3+}	0	10	$Al_2(SO_4)_3$	Essential for some plants (e.g. peas, maize, sunflowers, and cereals). Can be toxic to some plants below 10 ppm. Sometimes used to produce flower pigments (e.g. by Hydrangeas).
Silicon	Variable micronutrient	SiO_3^{2-}	0	140	K_2SiO_3, Na_2SiO_3, and H_2SiO_3	Present in most plants, abundant in cereal crops, grasses, and tree bark. Evidence that SiO_3^{2-} improves plant disease resistance exists.
Titanium	Variable micronutrient	Ti^{3+}	0	5	H_4TiO_4	Might be essential but trace Ti^{3+} is so ubiquitous that its addition is rarely warranted. At 5 ppm favorable growth effects in some crops are notable (e.g. pineapple and peas).

Cobalt	Non-essential micronutrient	Co^{2+}	0	0.1	$CoSO_4$	Required by rhizobia, important for legume root nodulation.
Sodium	Non-essential micronutrient	Na^+	0	Highly variable	Na_2SiO_3, Na_2SO_4, NaCl, $NaHCO_3$, and NaOH	Na^+ can partially replace K^+ in some plant functions but K^+ is still an essential nutrient.
Vanadium	Non-essential micronutrient	VO^{2+}	0	Trace, undetermined	$VOSO_4$	Beneficial for rhizobial N_2 fixation.
Lithium	Non-essential micronutrient	Li^+	0	Undetermined	Li_2SO_4, LiCl, and LiOH	Li^+ can increase the chlorophyll content of some plants (e.g. potato and pepper plants).

Organic Hydroponic Solutions

Organic fertilizers can be used to supplement or entirely replace the inorganic compounds used in conventional hydroponic solutions. However, using organic fertilizers introduces a number of challenges that are not easily resolved. Examples include:

- organic fertilizers are highly variable in their nutritional compositions. Even similar materials can differ significantly based on their source (e.g. the quality of manure varies based on an animal's diet).

- organic fertilizers are often sourced from animal byproducts, making disease transmission a serious concern for plants grown for human consumption or animal forage.

- organic fertilizers are often particulate and can clog substrates or other growing equipment. Sieving and/or milling the organic materials to fine dusts is often necessary.

- some organic materials (i.e. particularly manures and offal) can further degrade to emit foul odors.

Nevertheless, if precautions are taken, organic fertilizers can be used successfully in hydroponics.

Organically Sourced Macronutrients

Examples of suitable materials, with their average nutritional contents tabulated in terms of percent dried mass, are listed in the following table.

Organic material	N	P_2O_5	K_2O	CaO	MgO	SO_2	Comment
Bloodmeal	13.0%	2.0%	1.0%	0.5%	–	–	
Bone ashes	–	35.0%	–	46.0%	1.0%	0.5%	
Bonemeal	4.0%	22.5%	–	33.0%	0.5%	0.5%	
Hoof / Horn meal	14.0%	1.0%	–	2.5%	–	2.0%	
Fishmeal	9.5%	7.0%	–	0.5%	–	–	
Wool waste	3.5%	0.5%	2.0%	0.5%	–	–	
Wood ashes	–	2.0%	5.0%	33.0%	3.5%	1.0%	

Cottonseed ashes	–	5.5%	27.0%	9.5%	5.0%	2.5%	
Cottonseed meal	7.0%	3.0%	2.0%	0.5%	0.5%	–	
Dried locust or grass-hopper	10.0%	1.5%	0.5%	0.5%	–	–	
Leather waste	5.5% to 22%	–	–	–	–	–	Milled to a fine dust.
Kelp meal, liquid seaweed	1%	–	12%	–	–	–	Commercial products available.
Poultry manure	2% to 5%	2.5% to 3%	1.3% to 3%	4.0%	1.0%	2.0%	A liquid compost which is sieved to remove solids and checked for pathogens.
Sheep manure	2.0%	1.5%	3.0%	4.0%	2.0%	1.5%	Same as poultry manure.
Goat manure	1.5%	1.5%	3.0%	2.0%	–	–	Same as poultry manure.
Horse manure	3% to 6%	1.5%	2% to 5%	1.5%	1.0%	0.5%	Same as poultry manure.
Cow manure	2.0%	1.5%	2.0%	4.0%	1.1%	0.5%	Same as poultry manure.
Bat guano	8.0%	40%	29%	Trace	Trace	Trace	High in micronutrients. Commercially available.
Bird guano	13%	8%	20%	Trace	Trace	Trace	High in micronutrients. Commercially available.

Organically Sourced Micronutrients

Micronutrients can be sourced from organic fertilizers as well. For example, composted pine bark is high in manganese and is sometimes used to fulfill that mineral requirement in hydroponic solutions. To satisfy requirements for National Organic Programs, pulverized, unrefined minerals (e.g. Gypsum, Calcite, and glauconite) can also be added to satisfy a plant's nutritional needs.

Additives

In addition to chelating agents, humic acids can be added to increase nutrient uptake.

Tools

Common Equipment

Managing nutrient concentrations and pH values within acceptable ranges is essential for successful hydroponic horticulture. Common tools used to manage hydroponic solutions include:

- Electrical conductivity meters, a tool which estimates nutrient ppm by measuring how well a solution transmits an electric current.

- pH meter, a tool that uses an electric current to determine the concentration of hydrogen ions in solution.

- Litmus paper, disposable pH indicator strips that determine hydrogen ion concentrations by color changing chemical reaction.

- Graduated cylinders or measuring spoons to measure out premixed, commercial hydroponic solutions.

Advanced Equipment

Advanced equipment can also be used to perform accurate chemical analyses of nutrient solutions. Examples include:

- Balances for accurately measuring materials.

- Laboratory glassware, such as burettes and pipettes, for performing titrations.

- Colorimeters for solution tests which apply the Beer–Lambert law.

Using advanced equipment for hydroponic solutions can be beneficial to growers of any background because nutrient solutions are often reusable. Because nutrient solutions are virtually never completely depleted, and should never be due to the unacceptably low osmotic pressure that would result, re-fortification of old solutions with new nutrients can save growers money and can control point source pollution, a common source for the eutrophication of nearby lakes and streams.

Software

Although pre-mixed concentrated nutrient solutions are generally purchased from commercial nutrient manufacturers by hydroponic hobbyists and small commercial growers, several tools exist to help anyone prepare their own solutions without extensive knowledge about chemistry. The free and open source tools HydroBuddy and HydroCal have been created by professional chemists to help any hydroponics grower prepare their own nutrient solutions. The first program is available for Windows, Mac and Linux while the second one can be used through a simple JavaScript interface. Both programs allow for basic nutrient solution preparation although HydroBuddy provides added functionality to use and save custom substances, save formulations and predict electrical conductivity values.

Mixing Solutions

Often mixing hydroponic solutions using individual salts is impractical for hobbyists and/or small-scale commercial growers because commercial products are available at reasonable prices. However, even when buying commercial products, multi-component fertilizers are popular. Often these products are bought as three part formulas which emphasize certain nutritional roles. For example, solutions for vegetative growth (i.e. high in nitrogen), flowering (i.e. high in potassium and phosphorus), and micronutrient solutions (i.e. with trace minerals) are popular. The timing and application of these multi-part fertilizers should coincide with a plant's growth stage. For example, at the end of an annual plant's life cycle, a plant should be restricted from high nitrogen fertilizers. In most plants, nitrogen restriction inhibits vegetative growth and helps induce flowering.

Advancements

With pest problems reduced and nutrients constantly fed to the roots, productivity in hydroponics is high; however, growers can further increase yield by manipulating a plant's environment by constructing sophisticated growrooms.

CO_2 Enrichment

To increase yield further, some sealed greenhouses inject CO_2 into their environment to help improve growth and plant fertility.

References

- F. Stuart Chapin III; Pamela A. Matson; Harold A. Moon (2002). Principles of Terrestrial Ecosystem Ecology. Springer. ISBN 0387954392.

- Kötke, William H. (1993). The Final Empire: The Collapse of Civilization and the Seed of the Future. Arrow Point Press. ISBN 0963378457.

- Marschner, Petra, ed. (2012). Marschner's mineral nutrition of higher plants (3rd ed.). Amsterdam: Elsevier/ Academic Press. ISBN 9780123849052.

- Norman P. A. Huner; William Hopkins. "3 & 4". Introduction to Plant Physiology 4th Edition. John Wiley & Sons, Inc. ISBN 978-0-470-24766-2.

- Gericke, William F. (1940). The Complete Guide to Soilless Gardening (1st ed.). London: Putnam. pp. 9–10, 38 & 84. ISBN 9781163140499.

- Douglas, James Sholto (1975). Hydroponics: The Bengal System (5th ed.). New Dehli: Oxford University Press. p. 10. ISBN 9780195605662.

- Sholto Douglas, James (1985). Advanced guide to hydroponics: (soiless cultivation). London: Pelham Books. pp. 169–187, 289–320, & 345–351. ISBN 9780720715712.

- J. Benton, Jones (2004). Hydroponics: A Practical Guide for the Soilless Grower (2nd ed.). Newyork: Taylor & Francis. pp. 29–70 & 225–229. ISBN 9780849331671.

- Kumar, Ramasamy Rajesh; Cho, Jae Young (2014). "Reuse of hydroponic waste solution". Environmental Science and Pollution Research. 21 (16): 9569–9577. doi:10.1007/s11356-014-3024-3.

- Wallheimer, Brian (October 25, 2010). "Rice hulls a sustainable drainage option for greenhouse growers". Purdue University. Retrieved August 30, 2012.

- "AAPFCO Board of Directors 2006 Mid-Year Meeting" (PDF). Association of American Plant Food Control Officials. Retrieved 18 July 2011.

- Miranda, Stephen R.; Barker, Bruce (August 4, 2009). "Silicon: Summary of Extraction Methods". Harsco Minerals. Retrieved 18 July 2011.

Methods and Techniques of Crop Production

Recent advances in crop production techniques have led to bumper crops and high-yielding variety seeds. Such technologies enable farmers to gather greater output of crops during harvest while spending less time and energy. Tools and techniques are an important component of any field of study. The following chapter elucidates the various tools and techniques that are related to agricultural sciences.

Plant Genetics

An image of multiple chromosomes, making up a genome

Plant genetics is a very broad term. There are many facets of genetics in general, and of course there are many facets to plants. The definition of genetics is the branch of biology that deals with heredity, especially the mechanisms of hereditary transmission and the variation of inherited characteristics among similar or related organisms. And the definition of a plant is any of various photosynthetic, eukaryotic, multicellular organisms of the kingdom Plantae characteristically producing embryos, containing chloroplasts, having cell walls which contain cellulose, and lacking the power of locomotion. Although there has been a revolution in the biological sciences in the past twenty years, there is still a great deal that remains to be discovered. The completion of the sequencing of the genomes of rice and some agriculturally and scientifically important plants (for example Physcomitrella patens) has increased the possibilities of plant genetic research immeasurably.

Features of Plant Biology

Plant genetics is different from that of animals in a few ways. Like mitochondria, chloroplasts have their own DNA, complicating pedigrees somewhat. Like animals, plants have somatic mutations regularly, but these mutations can contribute to the germ line with ease, since flowers develop at the ends of branches composed of somatic cells. People have known of this for centuries, and mutant branches are called "sports". If the fruit on the sport is economically desirable, a new cultivar may be obtained.

Some plant species are capable of self-fertilization, and some are nearly exclusively self-fertilizers. This means that a plant can be both mother and father to its offspring, a rare occurrence in animals. Scientists and hobbyists attempting to make crosses between different plants must take special measures to prevent the plants from self-fertilizing.

Plants are generally more capable of surviving, and indeed flourishing, as polyploids. Polyploidy, the presence of extra sets of chromosomes, is not usually compatible with life in animals. In plants, polyploid individuals are created frequently by a variety of processes, and once created usually cannot cross back to the parental type. Polyploid individuals, if capable of self-fertilizing, can give rise to a new genetically distinct lineage, which can be the start of a new species. This is often called "instant speciation". Polyploids generally have larger fruit, an economically desirable trait, and many human food crops, including wheat, maize, potatoes, peanuts, strawberries and tobacco, are either accidentally or deliberately created polyploids.

Hybrids between plant species are easy to create by hand-pollination, and may be more successful on average than hybrids between animal species. Often tens of thousands of offspring from a single cross are raised and tested to obtain a single individual with desired characteristics. People create hybrids for economic and aesthetic reasons, especially with orchids.

DNA

Deoxyribonucleic acid (DNA) is a nucleic acid that contains the genetic instructions used in the development and functioning of all known living organisms and some viruses. The main role of DNA molecules is the long-term storage of information. DNA is often compared to a set of blueprints or a recipe, or a code, since it contains the instructions needed to construct other components of cells, such as proteins and RNA molecules. The DNA segments that carry this genetic information are called genes, but other DNA sequences have structural purposes, or are involved in regulating the use of this genetic information. Geneticists, including plant geneticists, use this sequencing of DNA to their advantage as they splice and delete certain genes and regions of the

DNA molecule to produce a different or desired genotype and thus, also producing a different phenotype.

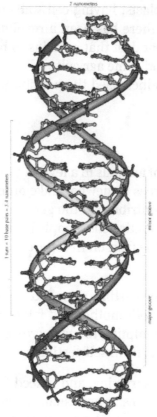

The structure of part of a DNA double helix

Gregor Mendel

Gregor Mendel was an Augustinian priest and scientist born on 20 July 1822 in Austria-Hungary and is well known for discovering genetics. He went to the Abbey of St. Thomas in Brno. He is often called the father of genetics for his study of the inheritance of certain traits in pea plants. Mendel showed that the inheritance of these traits follows particular laws, which were later named after him. The significance of Mendel's work was not recognized until the turn of the 20th century. Its rediscovery prompted the foundation of the discipline of genetics allows geneticists today to accurately predict the outcome of such crosses and in determining the phenotypical effects of the crosses. He died on 6 January 1884 from chronic nephritis.

Modern Ways to Genetically Modify Plants

There are two predominant procedures of transforming genes in organisms: the "Gene gun" method and the *Agrobacterium* method.

"Gene Gun" Method

The "Gene Gun" method is also referred to as "biolistics" (ballistics using biological components). This technique is used for in vivo (within a living organism) transformation and has been especially useful in transforming monocot species like corn and rice. This approach literally shoots genes into plant cells and plant cell chloroplasts. DNA is coated onto small particles of gold or tungsten approximately two micrometres in diameter. The particles are placed in a vacuum chamber and the plant tissue to be engineered is placed below the chamber. The particles are propelled at high velocity using a short pulse of high pressure helium gas, and hit a fine mesh baffle placed above the tissue while the DNA coating continues into any target cell or tissue.

Agrobacterium Method

Transformation via *Agrobacterium* has been successfully practiced in dicots, i.e. broadleaf plants, such as soybeans and tomatoes, for many years. Recently it has been adapted and is now effective in monocots like grasses, including corn and rice. In general, the *Agrobacterium* method is considered preferable to the gene gun, because of a greater frequency of single-site insertions of the foreign DNA, which allows for easier monitoring. In this method, the tumor inducing (Ti) region is removed from the T-DNA (transfer DNA) and replaced with the desired gene and a marker, which is then inserted into the organism. This may involve direct inoculation of the tissue with a culture of transformed Agrobacterium, or inoculation following treatment with micro-projectile bombardment, which wounds the tissue. Wounding of the target tissue causes the release of phenolic compounds by the plant, which induces invasion of the tissue by Agrobacterium. Because of this, microprojectile bombardment often increases the efficiency of infection with Agrobacterium. The marker is used to find the organism which has successfully taken up the desired gene. Tissues of the organism are then transferred to a medium containing an antibiotic or herbicide, depending on which marker was used. The *Agrobacterium* present is also killed by the antibiotic. Only tissues expressing the marker will survive and possess the gene of interest. Thus, subsequent steps in the process will only use these surviving plants. In order to obtain whole plants from these tissues, they are grown under controlled environmental conditions in tissue culture. This is a process of a series of media, each containing nutrients and hormones. Once the plants are grown and produce seed, the process of evaluating the progeny begins. This process entails selection of the seeds with the desired traits and then retesting and growing to make sure that the entire process has been completed successfully with the desired results.

Genetically Engineered Crops

The use of genetically engineered crops has helped many farmers deal with pest problems that reduce their crop production. The impact of pest-resistant crops has led to a much higher yield for farmers in today's world. They can use less pesticides which reduces the chemicals that they put into the ground. Certain engineered crops have led to farmers all over the world and in the United States to increase crop yield exponentially in recent years. Farmers can use a glyphosate herbicide to kill weeds, yet the genetically engineered corn is resistant to the herbicide and is left unaffected. Thus, fields are produced that are virtually weed free. Genetically engineered crops can also benefit farmers when dealing with potentially harmful viruses and bacteria. In the 1990s a mutant strain of virus was decimating the commercial corn fields of the United States. Scientists found a virus resistant strain of maize in the highlands of Mexico and extracted the part of the maize's genome that coded for resistance against the virus and incorporated it into their existing strain of commercial corn. This allowed the commercial strain to produce progeny that were resistant to the virus. Thus, the crops were saved from decimation.

Potential Detrimental Effects of Genetically Engineered Plants

According to John E. Berringer the outcome of releasing genetically modified organisms into the environment is still not known.

Genetically Modified Crops

Genetically modified crops (GMCs, GM crops, or biotech crops) are plants used in agriculture, the DNA of which has been modified using genetic engineering techniques. In most cases, the aim is to introduce a new trait to the plant which does not occur naturally in the species. Examples in food crops include resistance to certain pests, diseases, or environmental conditions, reduction of spoilage, or resistance to chemical treatments (e.g. resistance to a herbicide), or improving the nutrient profile of the crop. Examples in non-food crops include production of pharmaceutical agents, biofuels, and other industrially useful goods, as well as for bioremediation.

Farmers have widely adopted GM technology. Between 1996 and 2013, the total surface area of land cultivated with GM crops increased by a factor of 100, from 17,000 km^2 (4.2 million acres) to 1,750,000 km^2 (432 million acres). 10% of the world's arable land was planted with GM crops in 2010. In the US, by 2014, 94% of the planted area of soybeans, 96% of cotton and 93% of corn were genetically modified varieties. Use of GM crops expanded rapidly in developing countries, with about 18 million farmers growing 54% of worldwide GM crops by 2013. A 2014 meta-analysis concluded that GM technology adoption had reduced chemical pesticide use by 37%, increased crop yields by 22%, and increased farmer profits by 68%. This reduction in pesticide use has been ecologically beneficial, but benefits may be reduced by overuse. Yield gains and pesticide reductions are larger for insect-resistant crops than for herbicide-tolerant crops. Yield and profit gains are higher in developing countries than in developed countries.

There is a scientific consensus that currently available food derived from GM crops poses no greater risk to human health than conventional food, but that each GM food needs to be tested on a case-by-case basis before introduction. Nonetheless, members of the public are much less likely than scientists to perceive GM foods as safe. The legal and regulatory status of GM foods varies by country, with some nations banning or restricting them, and others permitting them with widely differing degrees of regulation.

However, opponents have objected to GM crops on several grounds, including environmental concerns, whether food produced from GM crops is safe, whether GM crops are needed to address the world's food needs, and concerns raised by the fact these organisms are subject to intellectual property law.

Gene transfer in nature and traditional agriculture

DNA transfers naturally between organisms. Several natural mechanisms allow gene flow across species. These occur in nature on a large scale – for example, it is one mechanism for the development of antibiotic resistance in bacteria. This is facilitated by transposons, retrotransposons, proviruses and other mobile genetic elements that naturally translocate DNA to new loci in a genome. Movement occurs over an evolutionary time scale.

The introduction of foreign germplasm into crops has been achieved by traditional crop breeders by overcoming species barriers. A hybrid cereal grain was created in 1875, by crossing wheat and rye. Since then important traits including dwarfing genes and rust resistance have been introduced. Plant tissue culture and deliberate mutations have enabled humans to alter the makeup of plant genomes.

History

The first genetically modified crop plant was produced in 1982, an antibiotic-resistant tobacco plant. The first field trials occurred in France and the USA in 1986, when tobacco plants were engineered for herbicide resistance. In 1987, Plant Genetic Systems (Ghent, Belgium), founded by Marc Van Montagu and Jeff Schell, was the first company to genetically engineer insect-resistant (tobacco) plants by incorporating genes that produced insecticidal proteins from *Bacillus thuringiensis* (Bt).

The People's Republic of China was the first country to allow commercialized transgenic plants, introducing a virus-resistant tobacco in 1992, which was withdrawn in 1997. The first genetically modified crop approved for sale in the U.S., in 1994, was the *FlavrSavr* tomato. It had a longer shelf life, because it took longer to soften after ripening. In 1994, the European Union approved tobacco engineered to be resistant to the herbicide bromoxynil, making it the first commercially genetically engineered crop marketed in Europe.

In 1995, Bt Potato was approved by the US Environmental Protection Agency, making it the country's first pesticide producing crop. In 1995 canola with modified oil composition (Calgene), Bt maize (Ciba-Geigy), bromoxynil-resistant cotton (Calgene), Bt cotton (Monsanto), glyphosate-resistant soybeans (Monsanto), virus-resistant squash (Asgrow), and additional delayed ripening tomatoes (DNAP, Zeneca/Peto, and Monsanto) were approved. As of mid-1996, a total of 35 approvals had been granted to commercially grow 8 transgenic crops and one flower crop (carnation), with 8 different traits in 6 countries plus the EU. In 2000, Vitamin A-enriched golden rice was developed, though as of 2016 it was not yet in commercial production. In 2013 the leaders of the three research teams that first applied genetic engineering to crops, Robert Fraley, Marc Van Montagu and Mary-Dell Chilton were awarded the World Food Prize for improving the "quality, quantity or availability" of food in the world.

Methods

Genetically engineered crops have genes added or removed using genetic engineering techniques, originally including gene guns, electroporation, microinjection and agrobacterium. More recently, CRISPR and TALEN offered much more precise and convenient editing techniques.

Gene guns (a.k.a. biolistic) "shoot" (direct high energy particles or radiations against) target genes into plant cells. It is the most common method. DNA is bound to tiny particles of gold or tungsten which are subsequently shot into plant tissue or single plant cells under high pressure. The accelerated particles penetrate both the cell wall and membranes. The DNA separates from the metal and is integrated into plant DNA inside the nucleus. This method has been applied successfully for many cultivated crops, especially monocots like wheat or maize, for which transformation using *Agrobacterium tumefaciens* has been less successful. The major disadvantage of this procedure is that serious damage can be done to the cellular tissue.

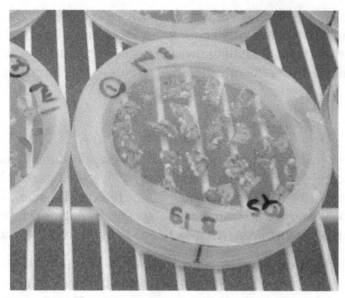

Plants (*Solanum chacoense*) being transformed using agrobacterium

Agrobacterium tumefaciens-mediated transformation is another common technique. Agrobacteria are natural plant parasites, and their natural ability to transfer genes provides another engineering method. To create a suitable environment for themselves, these Agrobacteria insert their genes into plant hosts, resulting in a proliferation of modified plant cells near the soil level (crown gall). The genetic information for tumour growth is encoded on a mobile, circular DNA fragment (plasmid). When Agrobacterium infects a plant, it transfers this T-DNA to a random site in the plant genome. When used in genetic engineering the bacterial T-DNA is removed from the bacterial plasmid and replaced with the desired foreign gene. The bacterium is a vector, enabling transportation of foreign genes into plants. This method works especially well for dicotyledonous plants like potatoes, tomatoes, and tobacco. Agrobacteria infection is less successful in crops like wheat and maize.

Electroporation is used when the plant tissue does not contain cell walls. In this technique, "DNA enters the plant cells through miniature pores which are temporarily caused by electric pulses."

Microinjection directly injects the gene into the DNA.

Plant scientists, backed by results of modern comprehensive profiling of crop composition, point out that crops modified using GM techniques are less likely to have unintended changes than are conventionally bred crops.

In research tobacco and *Arabidopsis thaliana* are the most frequently modified plants, due to well-developed transformation methods, easy propagation and well studied genomes. They serve as model organisms for other plant species.

Introducing new genes into plants requires a promoter specific to the area where the gene is to be expressed. For instance, to express a gene only in rice grains and not in leaves, an endosperm-specific promoter is used. The codons of the gene must be optimized for the organism due to codon usage bias. Transgenic gene products should be able to be denatured by heat so that they are destroyed during cooking.

Types of Modifications

Transgenic maize containing a gene from the bacteria *Bacillus thuringiensis*

Transgenic

Transgenic plants have genes inserted into them that are derived from another species. The inserted genes can come from species within the same kingdom (plant to plant) or between kingdoms (for example, bacteria to plant). In many cases the inserted DNA has to be modified slightly in order to correctly and efficiently express in the host organism. Transgenic plants are used to express proteins like the cry toxins from *B. thuringiensis*, herbicide resistant genes, antibodies and antigens for vaccinations A study led by the European Food Safety Authority (EFSA) found also viral genes in transgenic plants.

Transgenic carrots have been used to produce the drug Taliglucerase alfa which is used to treat Gaucher's disease. In the laboratory, transgenic plants have been modified to increase photosynthesis (currently about 2% at most plants versus the theoretic potential of 9–10%). This is possible by changing the rubisco enzyme (i.e. changing C3 plants into C4 plants), by placing the rubisco in a carboxysome, by adding CO_2 pumps in the cell wall, by changing the leaf form/size. Plants have been engineered to exhibit bioluminescence that may become a sustainable alternative to electric lighting.

Cisgenic

Cisgenic plants are made using genes found within the same species or a closely related one, where conventional plant breeding can occur. Some breeders and scientists argue that cisgenic modification is useful for plants that are difficult to crossbreed by conventional means (such as potatoes), and that plants in the cisgenic category should not require the same regulatory scrutiny as transgenics.

Subgenic

Genetically modified plants can also be developed using gene knockdown to alter the genetic makeup of a plant without incorporating genes from other plants. In 2014, Chinese researcher Gao Caixia filed patents on the creation of a strain of wheat that is resistant to powdery mildew. The strain lacks genes that encode proteins that repress defenses against the mildew. The researchers deleted all three copies of the genes from wheat's hexaploid genome. Gao used the TALENs and CRISPR gene editing tools without adding or changing any other genes. No field trials were immediately planned. The CRISPR technique has also been used to modify white button mushrooms (*Agaricus bisporus*).

Economics

GM food's economic value to farmers is one of its major benefits, including in developing nations. A 2010 study found that Bt corn provided economic benefits of $6.9 billion over the previous 14 years in five Midwestern states. The majority ($4.3 billion) accrued to farmers producing non-Bt corn. This was attributed to European corn borer populations reduced by exposure to Bt corn, leaving fewer to attack conventional corn nearby. Agriculture economists calculated that "world surplus [increased by] $240.3 million for 1996. Of this total, the largest share (59%) went to U.S. farmers. Seed company Monsanto received the next largest share (21%), followed by US consumers (9%), the rest of the world (6%), and the germplasm supplier, Delta & Pine Land Company of Mississippi (5%)."

According to the International Service for the Acquisition of Agri-biotech Applications (ISAAA), in 2014 approximately 18 million farmers grew biotech crops in 28 countries; about 94% of the farmers were resource-poor in developing countries. 53% of the global biotech crop area of 181.5 million hectares was grown in 20 developing countries. PG Economics comprehensive 2012 study concluded that GM crops increased farm incomes worldwide by $14 billion in 2010, with over half this total going to farmers in developing countries.

Critics challenged the claimed benefits to farmers over the prevalence of biased observers and by the absence of randomized controlled trials. The main Bt crop grown by small farmers in developing countries is cotton. A 2006 review of Bt cotton findings by agricultural economists concluded, "the overall balance sheet, though promising, is mixed. Economic returns are highly variable over years, farm type, and geographical location".

In 2013 the European Academies Science Advisory Council (EASAC) asked the EU to allow the development of agricultural GM technologies to enable more sustainable agriculture, by employing fewer land, water and nutrient resources. EASAC also criticizes the EU's "timeconsuming and expensive regulatory framework" and said that the EU had fallen behind in the adoption of GM technologies.

Participants in agriculture business markets include seed companies, agrochemical companies, distributors, farmers, grain elevators and universities that develop new crops/traits and whose agricultural extensions advise farmers on best practices. According to a 2012 review based on data from the late 1990s and early 2000s, much of the GM crop grown each year is used for livestock feed and increased demand for meat leads to increased demand for GM feedcrops. Feed grain usage as a percentage of total crop production is 70% for corn and more than 90% of oil seed meals such as soybeans. About 65 million metric tons of GM corn grains and about 70 million metric tons of soybean meals derived from GM soybean become feed.

In 2014 the global value of biotech seed was US\$15.7 billion; US\$11.3 billion (72%) was in industrial countries and US\$4.4 billion (28%) was in the developing countries. In 2009, Monsanto had \$7.3 billion in sales of seeds and from licensing its technology; DuPont, through its Pioneer subsidiary, was the next biggest company in that market. As of 2009, the overall Roundup line of products including the GM seeds represented about 50% of Monsanto's business.

Some patents on GM traits have expired, allowing the legal development of generic strains that include these traits. For example, generic glyphosate-tolerant GM soybean is now available. Another impact is that traits developed by one vendor can be added to another vendor's proprietary strains, potentially increasing product choice and competition. The patent on the first type of *Roundup Ready* crop that Monsanto produced (soybeans) expired in 2014 and the first harvest of off-patent soybeans occurs in the spring of 2015. Monsanto has broadly licensed the patent to other seed companies that include the glyphosate resistance trait in their seed products. About 150 companies have licensed the technology, including Syngenta and DuPont Pioneer.

Yield

In 2014, the largest review yet concluded that GM crops' effects on farming were positive. The meta-analysis considered all published English-language examinations of the agronomic and economic impacts between 1995 and March 2014 for three major GM crops: soybean, maize, and cotton. The study found that herbicide-tolerant crops have lower production costs, while for insect-resistant crops the reduced pesticide use was offset by higher seed prices, leaving overall production costs about the same.

Yields increased 9% for herbicide tolerance and 25% for insect resistant varieties. Farmers who adopted GM crops made 69% higher profits than those who did not. The review found that GM crops help farmers in developing countries, increasing yields by 14 percentage points.

The researchers considered some studies that were not peer-reviewed, and a few that did not report sample sizes. They attempted to correct for publication bias, by considering sources beyond academic journals. The large data set allowed the study to control for potentially confounding variables such as fertiliser use. Separately, they concluded that the funding source did not influence study results.

Traits

GM crops grown today, or under development, have been modified with various traits. These traits include improved shelf life, disease resistance, stress resistance, herbicide resistance, pest resistance, production of useful goods such as biofuel or drugs, and ability to absorb toxins and for use in bioremediation of pollution.

Recently, research and development has been targeted to enhancement of crops that are locally important in developing countries, such as insect-resistant cowpea for Africa and insect-resistant brinjal (eggplant).

Lifetime

The first genetically modified crop approved for sale in the U.S. was the *FlavrSavr* tomato, which had a longer shelf life. It is no longer on the market.

In November 2014, the USDA approved a GM potato that prevents bruising.

In February 2015 Arctic Apples were approved by the USDA, becoming the first genetically modified apple approved for US sale. Gene silencing was used to reduce the expression of polyphenol oxidase (PPO), thus preventing enzymatic browning of the fruit after it has been sliced open. The trait was added to Granny Smith and Golden Delicious varieties. The trait includes a bacterial antibiotic resistance gene that provides resistance to the antibiotic kanamycin. The genetic engineering involved cultivation in the presence of kanamycin, which allowed only resistant cultivars to survive. Humans consuming apples do not acquire kanamycin resistance, per arcticapple.com. The FDA approved the apples in March 2015.

Nutrition

Edible oils

Some GM soybeans offer improved oil profiles for processing or healthier eating. Camelina sativa has been modified to produce plants that accumulate high levels of oils similar to fish oils.

Vitamin Enrichment

Golden rice, developed by the International Rice Research Institute (IRRI), provides greater amounts of Vitamin A targeted at reducing Vitamin A deficiency. As of January 2016, golden rice has not yet been grown commercially in any country.

Researchers vitamin-enriched corn derived from South African white corn variety M37W, producing a 169-fold increase in Vitamin A, 6-fold increase in Vitamin C and doubled concentrations of folate. Modified Cavendish bananas express 10-fold the amount of Vitamin A as unmodified varieties.

Toxin Reduction

A genetically modified cassava under development offers lower cyanogen glucosides and enhanced protein and other nutrients (called BioCassava).

In November 2014, the USDA approved a potato, developed by J.R. Simplot Company, that prevents bruising and produces less acrylamide when fried. The modifications prevent natural, harmful proteins from being made via RNA interference. They do not employ genes from non-potato species. The trait was added to the Russet Burbank, Ranger Russet and Atlantic varieties.

Stress Resistance

Plants engineered to tolerate non-biological stressors such as drought, frost, high soil salinity, and nitrogen starvation were in development. In 2011, Monsanto's DroughtGard maize became the first drought-resistant GM crop to receive US marketing approval.

Herbicides

Glyphosate

As of 1999 the most prevalent GM trait was glyphosate-resistance. Glyphosate, (the active ingredient in Roundup and other herbicide products) kills plants by interfering with the shikimate pathway in plants, which is essential for the synthesis of the aromatic amino acids phenylalanine, tyrosine and tryptophan. The shikimate pathway is not present in animals, which instead obtain aromatic amino acids from their diet. More specifically, glyphosate inhibits the enzyme 5-enolpyruvylshikimate-3-phosphate synthase (EPSPS).

This trait was developed because the herbicides used on grain and grass crops at the time were highly toxic and not effective against narrow-leaved weeds. Thus, developing crops that could withstand spraying with glyphosate would both reduce environmental and health risks, and give an agricultural edge to the farmer.

Some micro-organisms have a version of EPSPS that is resistant to glyphosate inhibition. One of these was isolated from an *Agrobacterium* strain CP4 (CP4 EPSPS) that was resistant to glyphosate. The CP4 EPSPS gene was engineered for plant expression by fusing the 5' end of the gene to a chloroplast transit peptide derived from the petunia EPSPS. This transit peptide was used because it had shown previously an ability to deliver bacterial EPSPS to the chloroplasts of other plants. This CP4 EPSPS gene was cloned and transfected into soybeans.

The plasmid used to move the gene into soybeans was PV-GMGTO4. It contained three bacterial genes, two CP4 EPSPS genes, and a gene encoding beta-glucuronidase (GUS) from *Escherichia coli* as a marker. The DNA was injected into the soybeans using the particle acceleration method. Soybean cultivar A5403 was used for the transformation.

Bromoxynil

Tobacco plants have been engineered to be resistant to the herbicide bromoxynil.

Glufosinate

Crops have been commercialized that are resistant to the herbicide glufosinate, as well. Crops engineered for resistance to multiple herbicides to allow farmers to use a mixed group of two, three, or four different chemicals are under development to combat growing herbicide resistance.

2,4-D

In October 2014 the US EPA registered Dow's Enlist Duo maize, which is genetically modified to be resistant to both glyphosate and 2,4-D, in six states. Inserting a bacterial aryloxyalkanoate dioxygenase gene, *aad1* makes the corn resistant to 2,4-D. The USDA had approved maize and soybeans with the mutation in September 2014.

Dicamba

Monsanto has requested approval for a stacked strain that is tolerant of both glyphosate and dicamba.

Pest resistance

Insects

Tobacco, corn, rice and many other crops have been engineered to express genes encoding for insecticidal proteins from Bacillus thuringiensis (Bt). Papaya, potatoes, and squash have been engineered to resist viral pathogens such as cucumber mosaic virus which, despite its name, infects a wide variety of plants. The introduction of Bt crops during the period between 1996 and 2005 has been estimated to have reduced the total volume of insecticide active ingredient use in the United States by over 100 thousand tons. This represents a 19.4% reduction in insecticide use.

In the late 1990s, a genetically modified potato that was resistant to the Colorado potato beetle was withdrawn because major buyers rejected it, fearing consumer opposition.

Viruses

Virus resistant papaya were developed In response to a papaya ringspot virus (PRV) outbreak in Hawaii in the late 1990s. . They incorporate PRV DNA. By 2010, 80% of Hawaiian papaya plants were genetically modified.

Potatoes were engineered for resistance to potato leaf roll virus and Potato virus Y in 1998. Poor sales led to their market withdrawal after three years.

Yellow squash that were resistant to at first two, then three viruses were developed, beginning in the 1990s. The viruses are watermelon, cucumber and zucchini/courgette yellow mosaic. Squash was the second GM crop to be approved by US regulators. The trait was later added to zucchini.

Many strains of corn have been developed in recent years to combat the spread of Maize dwarf mosaic virus, a costly virus that causes stunted growth which is carried in Johnson grass and spread by aphid insect vectors. These strands are commercially available although the resistance is not standard among GM corn variants.

By-products

Drugs

In 2012, the FDA approved the first plant-produced pharmaceutical, a treatment for Gaucher's Disease. Tobacco plants have been modified to produce therapeutic antibodies.

Biofuel

Algae is under development for use in biofuels. Researchers in Singapore were working on GM jatropha for biofuel production. Syngenta has USDA approval to market a maize trademarked Enogen that has been genetically modified to convert its starch to sugar for ethanol. In 2013, the Flemish Institute for Biotechnology was investigating poplar trees genetically engineered to contain less lignin to ease conversion into ethanol. Lignin is the critical limiting factor when using wood to make bio-ethanol because lignin limits the accessibility of cellulose microfibrils to depolymerization by enzymes.

Materials

Companies and labs are working on plants that can be used to make bioplastics. Potatoes that produce industrially useful starches have been developed as well. Oilseed can be modified to produce fatty acids for detergents, substitute fuels and petrochemicals.

Bioremediation

Scientists at the University of York developed a weed (*Arabidopsis thaliana*) that contains genes from bacteria that can clean TNT and RDX-explosive soil contaminants. 16 million hectares in the USA (1.5% of the total surface) are estimated to be contaminated with TNT and RDX. However *A. thaliana* was not tough enough for use on military test grounds.

Genetically modified plants have been used for bioremediation of contaminated soils. Mercury, selenium and organic pollutants such as polychlorinated biphenyls (PCBs).

Marine environments are especially vulnerable since pollution such as oil spills are not containable. In addition to anthropogenic pollution, millions of tons of petroleum annually enter the marine environment from natural seepages. Despite its toxicity, a considerable fraction of petroleum oil entering marine systems is eliminated by the hydrocarbon-degrading activities of microbial communities. Particularly successful is a recently discovered group of specialists, the so-called hydrocarbonoclastic bacteria (HCCB) that may offer useful genes.

Asexual reproduction

Crops such as maize reproduce sexually each year. This randomizes which genes get propagated to the next generation, meaning that desirable traits can be lost. To maintain a high-quality crop, some farmers purchase seeds every year. Typically, the seed company maintains two inbred varieties, and crosses them into a hybrid strain that is then sold. Related plants like sorghum and gamma grass are able to perform apomixis, a form of asexual reproduction that keeps the plant's DNA intact. This trait is apparently controlled by a single dominant gene, but traditional breeding has been unsuccessful in creating asexually-reproducing maize. Genetic engineering offers another route to this goal. Successful modification would allow farmers to replant harvested seeds that retain desirable traits, rather than relying on purchased seed.

Crops

As of 2010 food species for which a genetically modified version is being commercially grown (percent modified in the table below are mostly 2009/2010 data) include:

Crop	Traits	Modification	Percent modified in US	Percent modified in world
Alfalfa	Tolerance of glyphosate or glufosinate	Genes added	Planted in the US from 2005–2007; 2007–2010 court injunction; 2011 approved for sale	

Apples	Delayed browning	Genes added for reduced polyphenol oxidase (PPO) production from other apples	2015 approved for sale	
Canola/ Rapeseed	Tolerance of glyphosate or glufosinate High laurate canola, Oleic acid canola	Genes added	87% (2005)	21%
Corn	Tolerance of herbicides glyphosate glufosinate, and 2,4-D. Insect resistance. Added enzyme, alpha amylase, that converts starch into sugar to facilitate ethanol production. Viral resistance	Genes, some from Bt, added.	Herbicide-resistant: 2013, 85% Bt: 2013, 76% Stacked: 2013, 71%	26%
Cotton (cottonseed oil)	Insect resistance	Genes, some from Bt, added	Herbicide-resistant: 2013, 82% Bt: 2013, 75% Stacked: 2013, 71%	49%
Eggplant	Insect resistance	Genes from Bt	Negligible	Negligible
Papaya (Hawaiian)	Resistance to the papaya ringspot virus.	Gene added	80%	
Potato (food)	Resistance to Colorado beetle Resistance to potato leaf roll virus and Potato virus Y Reduced acrylamide when fried and reduced bruising	Bt cry3A, coat protein from PVY "Innate" potatoes added genetic material coding for mRNA for RNA interference	0%	0%
Potato (starch)	Antibiotic resistance gene, used for selection Better starch production	Antibiotic resistance gene from bacteria Modifications to endogenous starch-producing enzymes	0%	0%
Rice	Enriched with beta-carotene (a source of vitamin A)	Genes from maize and a common soil microorganism.		
Soybeans	Tolerance of glyphosate or glufosinate Reduced saturated fats (high oleic acid); Kills susceptible insect pests Viral resistance	Herbicide resistant gene taken from bacteria added Knocked out native genes that catalyze saturation Gene for one or more Bt crystal proteins added	2014: 94%	77%
Squash (Zucchini/ Courgette)	Resistance to watermelon, cucumber and zucchini/courgette yellow mosaic viruses	Viral coat protein genes	13% (figure is from 2005)	
Sugar beet	Tolerance of glyphosate, glufosinate	Genes added	95% (2010); regulated 2011; deregulated 2012	9%
Sugarcane	Pesticide tolerance High sucrose content.	Genes added		

Sweet peppers	Resistance to cucumber mosaic virus	Viral coat protein genes		Small quantities grown in China
Tomatoes	Suppression of the enzyme polygalacturonase (PG), retarding fruit softening after harvesting, while at the same time retaining both the natural color and flavor of the fruit	Antisense gene of the gene responsible for PG enzyme production added	Taken off the market due to commercial failure.	Small quantities grown in China

Development

The number of USDA-approved field releases for testing grew from 4 in 1985 to 1,194 in 2002 and averaged around 800 per year thereafter. The number of sites per release and the number of gene constructs (ways that the gene of interest is packaged together with other elements)—have rapidly increased since 2005. Releases with agronomic properties (such as drought resistance) jumped from 1,043 in 2005 to 5,190 in 2013. As of September 2013, about 7,800 releases had been approved for corn, more than 2,200 for soybeans, more than 1,100 for cotton, and about 900 for potatoes. Releases were approved for herbicide tolerance (6,772 releases), insect resistance (4,809), product quality such as flavor or nutrition (4,896), agronomic properties like drought resistance (5,190), and virus/fungal resistance (2,616). The institutions with the most authorized field releases include Monsanto with 6,782, Pioneer/DuPont with 1,405, Syngenta with 565, and USDA's Agricultural Research Service with 370. As of September 2013 USDA had received proposals for releasing GM rice, squash, plum, rose, tobacco, flax and chicory.

Farming Practices

Bt Resistance

Constant exposure to a toxin creates evolutionary pressure for pests resistant to that toxin. Over-reliance on glyphosate and a reduction in the diversity of weed management practices allowed the spread of glyphosate resistance in 14 weed species/biotypes in the US.

One method of reducing resistance is the creation of refuges to allow nonresistant organisms to survive and maintain a susceptible population.

To reduce resistance to Bt crops, the 1996 commercialization of transgenic cotton and maize came with a management strategy to prevent insects from becoming resistant. Insect resistance management plans are mandatory for Bt crops. The aim is to encourage a large population of pests so that any (recessive) resistance genes are diluted within the population. Resistance lowers evolutionary fitness in the absence of the stressor (Bt). In refuges, non-resistant strains outcompete resistant ones.

With sufficiently high levels of transgene expression, nearly all of the heterozygotes (S/s), i.e., the largest segment of the pest population carrying a resistance allele, will be killed before maturation, thus preventing transmission of the resistance gene to their progeny. Refuges (i. e., fields of non-transgenic plants) adjacent to transgenic fields increases the likelihood that homozygous resistant (s/s) individuals and any surviving heterozygotes will mate with susceptible (S/S) individuals from the refuge, instead of with other individuals carrying the resistance allele. As a result, the resistance gene frequency in the population remains lower.

Complicating factors can affect the success of the high-dose/refuge strategy. For example, if the temperature is not ideal, thermal stress can lower Bt toxin production and leave the plant more susceptible. More importantly, reduced late-season expression has been documented, possibly resulting from DNA methylation of the promoter. The success of the high-dose/refuge strategy has successfully maintained the value of Bt crops. This success has depended on factors independent of management strategy, including low initial resistance allele frequencies, fitness costs associated with resistance, and the abundance of non-Bt host plants outside the refuges.

Companies that produce Bt seed are introducing strains with multiple Bt proteins. Monsanto did this with Bt cotton in India, where the product was rapidly adopted. Monsanto has also; in an attempt to simplify the process of implementing refuges in fields to comply with Insect Resistance Management(IRM) policies and prevent irresponsible planting practices; begun marketing seed bags with a set proportion of refuge (non-transgenic) seeds mixed in with the Bt seeds being sold. Coined "Refuge-In-a-Bag" (RIB), this practice is intended to increase farmer compliance with refuge requirements and reduce additional labor needed at planting from having separate Bt and refuge seed bags on hand. This strategy is likely to reduce the likelihood of Bt-resistance occurring for corn rootworm, but may increase the risk of resistance for lepidopteran corn pests, such as European corn borer. Increased concerns for resistance with seed mixtures include partially resistant larvae on a Bt plant being able to move to a susceptible plant to survive or cross pollination of refuge pollen on to Bt plants that can lower the amount of Bt expressed in kernels for ear feeding insects.

Herbicide Resistance

Best management practices (BMPs) to control weeds may help delay resistance. BMPs include applying multiple herbicides with different modes of action, rotating crops, planting weed-free seed, scouting fields routinely, cleaning equipment to reduce the transmission of weeds to other fields, and maintaining field borders. The most widely planted GMOs are designed to tolerate herbicides. By 2006 some weed populations had evolved to tolerate some of the same herbicides. Palmer amaranth is a weed that competes with cotton. A native of the southwestern US, it traveled east and was first found resistant to glyphosate in 2006, less than 10 years after GM cotton was introduced.

Plant Protection

Farmers generally use less insecticide when they plant Bt-resistant crops. Insecticide use on corn farms declined from 0.21 pound per planted acre in 1995 to 0.02 pound in 2010. This is consistent with the decline in European corn borer populations as a direct result of Bt corn and cotton. The establishment of minimum refuge requirements helped delay the evolution of Bt resistance. However resistance appears to be developing to some Bt traits in some areas.

Tillage

By leaving at least 30% of crop residue on the soil surface from harvest through planting, conservation tillage reduces soil erosion from wind and water, increases water retention, and reduces soil degradation as well as water and chemical runoff. In addition, conservation tillage reduces the carbon footprint of agriculture. A 2014 review covering 12 states from 1996 to 2006, found that a 1% increase in herbicde-tolerant (HT) soybean adoption leads to a 0.21% increase in conservation tillage and a 0.3% decrease in quality-adjusted herbicide use.

Regulation

The regulation of genetic engineering concerns the approaches taken by governments to assess and manage the risks associated with the development and release of genetically modified crops. There are differences in the regulation of GM crops between countries, with some of the most marked differences occurring between the USA and Europe. Regulation varies in a given country depending on the intended use of each product. For example, a crop not intended for food use is generally not reviewed by authorities responsible for food safety.

Production

In 2013, GM crops were planted in 27 countries; 19 were developing countries and 8 were developed countries. 2013 was the second year in which developing countries grew a majority (54%) of the total GM harvest. 18 million farmers grew GM crops; around 90% were small-holding farmers in developing countries.

Country	2013– GM planted area (million hectares)	Biotech crops
USA	70.1	Maize, Soybean, Cotton, Canola, Sugarbeet, Alfalfa, Papaya, Squash
Brazil	40.3	Soybean, Maize, Cotton
Argentina	24.4	Soybean, Maize, Cotton
India	11.0	Cotton
Canada	10.8	Canola, Maize, Soybean, Sugarbeet
Total	175.2	----

The United States Department of Agriculture (USDA) reports every year on the total area of GMO varieties planted in the United States. According to National Agricultural Statistics Service, the states published in these tables represent 81–86 percent of all corn planted area, 88–90 percent of all soybean planted area, and 81–93 percent of all upland cotton planted area (depending on the year).

Global estimates are produced by the International Service for the Acquisition of Agri-biotech Applications (ISAAA) and can be found in their annual reports, "Global Status of Commercialized Transgenic Crops".

Farmers have widely adopted GM technology. Between 1996 and 2013, the total surface area of land cultivated with GM crops increased by a factor of 100, from 17,000 square kilometers (4,200,000 acres) to 1,750,000 km² (432 million acres). 10% of the world's arable land was planted with GM crops in 2010. As of 2011, 11 different transgenic crops were grown commercially on 395 million acres (160 million hectares) in 29 countries such as the USA, Brazil, Argentina, India, Canada, China, Paraguay, Pakistan, South Africa, Uruguay, Bolivia, Australia, Philippines, Myanmar, Burkina Faso, Mexico and Spain. One of the key reasons for this widespread adoption is the perceived economic benefit the technology brings to farmers. For example, the system of planting glyphosate-resistant seed and then applying glyphosate once plants emerged provided farmers with the opportunity to dramatically increase the yield from a given plot of land, since this allowed them to plant rows closer together. Without it, farmers had to plant rows far enough apart to control post-emergent weeds with mechanical tillage. Likewise, using Bt seeds means that farmers do not have to purchase insecticides, and then invest time, fuel, and equipment in applying them. However critics have disputed whether yields are higher and whether chemical use is less, with GM crops.

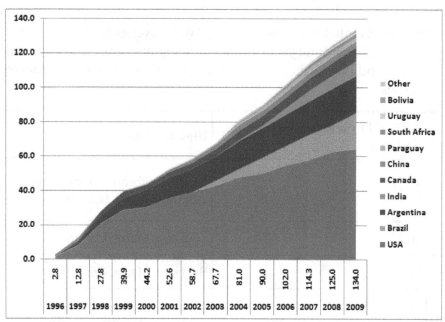

Land area used for genetically modified crops by country (1996–2009), in millions of hectares. In 2011, the land area used was 160 million hectares, or 1.6 million square kilometers.

In the US, by 2014, 94% of the planted area of soybeans, 96% of cotton and 93% of corn were genetically modified varieties. Genetically modified soybeans carried herbicide-tolerant traits only, but maize and cotton carried both herbicide tolerance and insect protection traits (the latter largely Bt protein). These constitute "input-traits" that are aimed to financially benefit the producers, but may have indirect environmental benefits and cost benefits to consumers. The Grocery Manufacturers of America estimated in 2003 that 70–75% of all processed foods in the U.S. contained a GM ingredient.

Europe grows relatively few genetically engineered crops with the exception of Spain, where one fifth of maize is genetically engineered, and smaller amounts in five other countries. The EU had a 'de facto' ban on the approval of new GM crops, from 1999 until 2004. GM crops are now regulated by the EU. In 2015, genetically engineered crops are banned in 38 countries worldwide, 19 of them in Europe. Developing countries grew 54 percent of genetically engineered crops in 2013.

In recent years GM crops expanded rapidly in developing countries. In 2013 approximately 18 million farmers grew 54% of worldwide GM crops in developing countries. 2013's largest increase was in Brazil (403,000 km² versus 368,000 km² in 2012). GM cotton began growing in India in 2002, reaching 110,000 km² in 2013.

According to the 2013 ISAAA brief: "...a total of 36 countries (35 + EU-28) have granted regulatory approvals for biotech crops for food and/or feed use and for environmental release or planting since 1994... a total of 2,833 regulatory approvals involving 27 GM crops and 336 GM events (NB: an "event" is a specific genetic modification in a specific species) have been issued by authorities, of which 1,321 are for food use (direct use or processing), 918 for feed use (direct use or processing) and 599 for environmental release or planting. Japan has the largest number (198), followed by the U.S.A. (165, not including "stacked" events), Canada (146), Mexico (131), South Korea (103), Australia (93), New Zealand (83), European Union (71 including approvals that have expired or under renewal process), Philippines (68), Taiwan (65), Colombia (59), China (55) and South Africa (52). Maize has the largest number (130 events in 27 countries), followed by cotton (49 events in 22 countries), potato (31 events in 10 countries), canola (30 events in 12 countries) and soybean (27 events in 26 countries).

Controversy

GM foods are controversial and the subject of protests, vandalism, referenda, legislation, court action and scientific disputes. The controversies involve consumers, biotechnology companies, governmental regulators, non-governmental organizations and scientists. The key areas are whether GM food should be labeled, the role of government regulators, the effect of GM crops on health and the environment, the effects of pesticide use and resistance, the impact on farmers, and their roles in feeding the world and energy production.

There is a scientific consensus that currently available food derived from GM crops poses no greater risk to human health than conventional food, but that each GM food needs to be tested on a case-by-case basis before introduction. Nonetheless, members of the public are much less likely than scientists to perceive GM foods as safe. The legal and regulatory status of GM foods varies by country, with some nations banning or restricting them, and others permitting them with widely differing degrees of regulation.

No reports of ill effects have been documented in the human population from GM food. Although GMO labeling is required in many countries, the United States Food and Drug Administration does not require labeling, nor does it recognize a distinction between approved GMO and non-GMO foods.

Advocacy groups such as Center for Food Safety, Union of Concerned Scientists, Greenpeace and the World Wildlife Fund claim that risks related to GM food have not been adequately examined and managed, that GMOs are not sufficiently tested and should be labelled, and that regulatory authorities and scientific bodies are too closely tied to industry. Some studies have claimed that genetically modified crops can cause harm; a 2016 review that reanalyzed the data from six of these studies found that their statistical methodologies were flawed and did not demonstrate harm, and said that conclusions about GMO crop safety should be drawn from "the totality of the evidence... instead of far-fetched evidence from single studies".

Examples Of Genetically Modified Crops

Genetically Modified Soybean

A genetically modified soybean is a soybean (*Glycine max*) that has had DNA introduced into it using genetic engineering techniques. In 1994 the first genetically modified soybean was introduced to the U.S. market, by Monsanto. In 2014, 90.7 million hectares of GM soy were planted worldwide, 82% of the total soy cultivation area.

Examples of Transgenic Soybeans

The genetic makeup of a soybean gives it a wide variety of uses, thus keeping it in high demand. First, manufacturers only wanted to use transgenics to be able to grow more soy at a minimal cost to meet this demand, and to fix any problems in the growing process, but they eventually found they could modify the soybean to contain healthier components, or even focus on one aspect of the soybean to produce in larger quantities. These phases became known as the first and second generation of genetically modified (GM) foods. As Peter Celec describes, "benefits of the first generation of GM foods were oriented towards the production process and companies, the second generation of GM foods offers, on contrary, various advantages and added value for the consumer", including "improved nutritional composition or even therapeutic effects."

Roundup Ready Soybean

Roundup Ready Soybeans (The first variety was also known as GTS 40-3-2 (OECD UI: MON-04032-6)) are a series of genetically engineered varieties of glyphosate-resistant soybeans produced by Monsanto.

Glyphosate kills plants by interfering with the synthesis of the essential amino acids phenylalanine, tyrosine and tryptophan. These amino acids are called "essential" because animals cannot make them; only plants and micro-organisms can make them and animals obtain them by eating plants.

Plants and microorganisms make these amino acids with an enzyme that only plants and lower organisms have, called 5-enolpyruvylshikimate-3-phosphate synthase (EPSPS). EPSPS is not present in animals, which instead obtain aromatic amino acids from their diet.

Roundup Ready Soybeans express a version of EPSPS from the CP4 strain of the bacteria, *Agrobacterium tumefaciens*, expression of which is regulated by an enhanced 35S promoter (E35S) from cauliflower mosaic virus (CaMV), a chloroplast transit peptide (CTP4) coding sequence from Petunia hybrida, and a nopaline synthase (nos 3') transcriptional termination element from Agrobacterium tumefaciens. The plasmid with EPSPS and the other genetic elements mentioned above was inserted into soybean germplasm with a gene gun by scientists at Monsanto and Asgrow. The patent on the first generation of Roundup Ready soybeans expired in March 2015.

History

First approved commercially in the United States during 1994, GTS 40-3-2 was subsequently introduced to Canada in 1995, Japan and Argentina in 1996, Uruguay in 1997, Mexico and Brazil in 1998, and South Africa in 2001.

Detection

GTS 40-3-2 can be detected using both nucleic acid and protein analysis methods.

Generic GMO Soybeans

Following expiration of Monsanto's patent on the first variety of glyphosate-resistant Roundup Ready soybeans, development began on glyphosate-resistant "generic" soybeans. The first variety, developed at the University of Arkansas Division of Agriculture, came on the market in 2015. With a slightly lower yield than newer Monsanto varieties, it costs about half as much, and seeds can be saved for subsequent years. According to its creator it is adapted to conditions in Arkansas. Several other varieties are being bred by crossing the original variety of Roundup Ready soybeans with other soybean varieties.

Stacked Traits

Monsanto developed a glyphosate-resistant soybean that also expresses Cry1Ac protein from Bacillus thuringiensis and the glyphosate-resistance gene, which completed the Brazilian regulatory process in 2010.

Genetic Modification to Improve Soybean Oil

Soy has been genetically modified to improve the quality of soy oil. Soy oil has a fatty acid profile that makes it susceptible to oxidation, which makes it rancid, and this has limited its usefulness to the food industry. Genetic modifications increased the amount of oleic acid and stearic acid and decreased the amount of linolenic acid. By silencing, or knocking out, the delta 9 and delta 12 desaturases. DuPont Pioneer created a high oleic fatty acid soybean with levels of oleic acid greater than 80%, and started marketing it in 2010.

Regulation

The regulation of genetic engineering concerns the approaches taken by governments to assess and manage the risks associated with the development and release of genetically modified crops. There are differences in the regulation of GM crops between countries, with some of the most marked differences occurring between the USA and Europe. Regulation varies in a given country depending on the intended use of the products of the genetic engineering. For example, a crop not intended for food use is generally not reviewed by authorities responsible for food safety.

Controversy

There is a scientific consensus that currently available food derived from GM crops poses no greater risk to human health than conventional food, but that each GM food needs to be tested on a case-by-case basis before introduction. Nonetheless, members of the public are much less likely than scientists to perceive GM foods as safe. The legal and regulatory status of GM foods varies by country, with some nations banning or restricting them, and others permitting them with widely differing degrees of regulation.

A 2010 study found that in the United States, GM crops also provide a number of ecological benefits.

Critics have objected to GM crops on several grounds, including ecological concerns, and economic concerns raised by the fact these organisms are subject to intellectual property law. GM crops also are involved in controversies over GM food with respect to whether food produced from GM crops is safe and whether GM crops are needed to address the world's food needs. These controversies have led to litigation, international trade disputes, and protests, and to restrictive legislation in most countries.

Gene Gun

PDS-1000/He Particle Delivery System

A gene gun or a biolistic particle delivery system, originally designed for plant transformation, is a device for injecting cells with genetic information; the inserted genetic material are termed transgenes. The payload is an elemental particle of a heavy metal coated with plasmid DNA. This technique is often simply referred to as bioballistics or biolistics.

This device is able to transform almost any type of cell, including plants, and is not limited to genetic material of the nucleus: it can also transform organelles, including plastids.

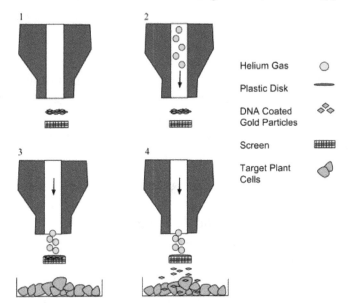

A gene gun is used for injecting cells with genetic information, it is also known as biolistic particle delivery system. Gene guns can be used effectively on most cells but are mainly used on plant cells. Step 1 The gene gun apparatus is ready to fire. Step 2 When the gun is turned on and the helium flows through. Step 3 The helium moving the disk with DNA coated particles toward the screen. Step 4 The helium having pushed the particles moving through the screen and moving to the target cells to transform the cells.

Gene Gun Design

The gene gun was originally a Crosman air pistol modified to fire dense tungsten particles. It was invented by John C Sanford, Ed Wolf and Nelson Allen at Cornell University, and Ted Klein of DuPont, between 1983 and 1986. The original target was onions (chosen for their large cell size) and it was used to deliver particles coated with a marker gene. Genetic transformation was then proven when the onion tissue expressed the gene.

The earliest custom manufactured gene guns (fabricated by Nelson Allen) used a 22 caliber nail gun cartridge to propel an extruded polyethylene cylinder (bullet) down a 22 cal. Douglas barrel. A droplet of the tungsten powder and genetic material was placed on the bullet and shot down the barrel at a lexan "stopping" disk with a petri dish below. The bullet welded to the disk and the genetic information blasted into the sample in the dish with a doughnut effect (devastation in the middle, a ring of good transformation and little around the edge). The gun was connected to a vacuum pump and was under vacuum while firing. The early design was put into limited production by a Rumsey-Loomis (a local machine shop then at Mecklenburg Rd in Ithaca, NY, USA). Later the design was refined by removing the "surge tank" and changing to nonexplosive propellants. DuPont added a plastic extrusion to the exterior to visually improve the machine for mass production to the scientific community. Biorad contracted with Dupont to manufacture and distribute the device. Improvements include the use of helium propellant and a multi-disk-collision delivery mechanism. Other heavy metals such as gold and silver are also used. Gold may be favored because it has better uniformity than tungsten and tungsten can be toxic to cells, but its use may be limited due to availability and cost.

Biolistic Construct Design

A construct is a piece of DNA inserted into the target's genome, including parts that are intended to be removed later. All biolistic transformations require a construct to proceed and while there is great variation among biolistic constructs, they can be broadly sorted into two categories: those which are designed to transform eukaryotic nuclei, and those designed to transform prokaryotic-type genomes such as mitochondria, plasmids or plastids.

Those meant to transform prokaryotic genomes generally have the gene or genes of interest, at least one promoter and terminator sequence, and a reporter gene; which is a gene used to ease detection or removal of those cells which didn't integrate the construct into their DNA. These genes may each have their own promoter and terminator, or be grouped to produce multiple gene products from one transcript, in which case binding sites for translational machinery should be placed between each to ensure maximum translational efficiency. In any case the entire construct is flanked by regions called border sequences which are similar in sequence to locations within the genome, this allows the construct to target itself to a specific point in the existing genome.

Constructs meant for integration into a eukaryotic nucleus follow a similar pattern except that: the construct contains no border sequences because the sequence rearrangement that prokaryotic constructs rely on rarely occurs in eukaryotes; and each gene contained within the construct must be expressed by its own copy of a promoter and terminator sequence.

Though the above designs are generally followed, there are exceptions. For example, the construct might include a Cre-Lox system to selectively remove inserted genes; or a prokaryotic construct may insert itself downstream of a promoter, allowing the inserted genes to be governed by a promoter already in place and eliminating the need for one to be included in the construct.

Application

Gene guns are so far mostly applied for plant cells. However, there is much potential use in humans and other animals as well.

Plants

The target of a gene gun is often a callus of undifferentiated plant cells growing on gel medium in a Petri dish. After the gold particles have impacted the dish, the gel and callus are largely disrupted. However, some cells were not obliterated in the impact, and have successfully enveloped a DNA coated gold particle, whose DNA eventually migrates to and integrates into a plant chromosome.

Cells from the entire Petri dish can be re-collected and selected for successful integration and expression of new DNA using modern biochemical techniques, such as a using a tandem selectable gene and northern blots.

Selected single cells from the callus can be treated with a series of plant hormones, such as auxins and gibberellins, and each may divide and differentiate into the organized, specialized, tissue cells of an entire plant. This capability of total re-generation is called totipotency. The new plant that originated from a successfully shot cell may have new genetic (heritable) traits.

The use of the gene gun may be contrasted with the use of *Agrobacterium tumefaciens* and its Ti plasmid to insert genetic information into plant cells.

Humans and Other Animals

Gene guns have also been used to deliver DNA vaccines.

The delivery of plasmids into rat neurons through the use of a gene gun, specifically DRG neurons, is also used as a pharmacological precursor in studying the effects of neurodegenerative diseases such as Alzheimer's disease.

The gene gun has become a common tool for labeling subsets of cells in cultured tissue. In addition to being able to transfect cells with DNA plasmids coding for fluorescent proteins, the gene gun can be adapted to deliver a wide variety of vital dyes to cells.

Gene gun bombardment has also been used to transform *Caenorhabditis elegans*, as an alternative to microinjection.

Advantages

Biolistics has proven to be a versatile method of genetic modification and it is generally preferred to engineer transformation-resistant crops, such as cereals. Notably, *Bt* maize is a product of biolistics. Plastid transformation has also seen great success with particle bombardment when compared to other current techniques, such as *Agrobacterium* mediated transformation, which have difficulty targeting the vector to and stably expressing in the chloroplast. In addition, there are no reports of a chloroplast silencing a transgene inserted with a gene gun. Additionally, with only one firing of a gene gun, a skilled technician can generate two transformed organisms. This technology has even allowed for modification of specific tissues *in situ*, although this is likely to damage large numbers of cells and transform only some, rather than all, cells of the tissue.

Limitations

However, biolistics introduces the construct randomly into the target cells. Thus the altered DNA sequences may be transformed into whatever genomes are present in the cell, be they nuclear, mitochondrial, plasmid or any others, in any combination, though proper construct design may mitigate this. Another issue is that the gene inserted may be overexpressed when the construct is inserted multiple times in either the same or different locations of the genome. This is due to the ability of the constructs to give and take genetic information from other constructs, causing some to carry no transgene and others to carry multiple copies; the number of copies inserted depends on both how many copies of the transgene an inserted construct has, and how many were inserted. Also, because eukaryotic constructs rely on illegitimate recombination, a process by which the transgene is integrated into the genome without similar genetic sequences, and not homologous recombination, which inserts at similar sequences, they cannot be targeted to specific locations within the genome.

References

- Martins VAP (2008). "Genomic Insights into Oil Biodegradation in Marine Systems". Microbial Biodegradation: Genomics and Molecular Biology. Caister Academic Press. ISBN 978-1-904455-17-2.

- "A decade of EU-funded GMO research (2001–2010)" (PDF). Directorate-General for Research and Innovation. Biotechnologies, Agriculture, Food. European Commission, European Union. 2010. doi:10.2777/97784. ISBN 978-92-79-16344-9. Retrieved February 8, 2016.

- Slater, Adrian; Scott, Nigel; Fowler, Mark (2008). Plant Biotechnology: the genetic manipulation of plants (2 ed.). Oxford, New York, USA: Oxford University Press Inc. ISBN 978-0-19-928261-6.

- Praitis, Vida (2006). "Creation of Transgenic Lines Using Microparticle Bombardment Methods" 351: 93–108. doi:10.1385/1-59745-151-7:93. ISBN 1-59745-151-7.

- Hayward, M.D.; Bosemark, N.O.; Romagosa, T. (2012). Plant Breeding: Principles and Prospects. Springer Science & Business Media. p. 131. ISBN 9789401115247.

- Lynch, Diahanna; Vogel, David (April 5, 2001). "The Regulation of GMOs in Europe and the United States: A Case-Study of Contemporary European Regulatory Politics". Council on Foreign Relations. Retrieved February 24, 2016.

- Paull, John (2015) The threat of genetically modified organisms (GMOs) to organic agriculture: A case study update, Agriculture & Food, 3: 56-63.

- Krimsky, Sheldon (2015). "An Illusory Consensus behind GMO Health Assessment" (PDF). Science, Technology, & Human Values. 40: 1–32. doi:10.1177/0162243915598381.

- Panchin, Alexander Y.; Tuzhikov, Alexander I. (2016). "Published GMO studies find no evidence of harm when corrected for multiple comparisons". Critical Reviews in Biotechnology: 1–5. doi:10.3109/07388551.2015.113 0684. PMID 26767435

- Fred Miller, University of Arkansas Division of Agriculture Communications (December 3, 2014). "Arkansas: 'Look Ma, No Tech Fees.' Round Up Ready Soybean Variety Released". AGFAX. Retrieved July 30, 2015. Monsanto's patent on the first generation of Roundup Ready products expires in March 2015....

- Antonio Regalado (July 30, 2015). "Monsanto no longer controls one of the biggest innovations in the history of agriculture.". MIT Technology Review. Retrieved July 30, 2015.

- Sanford, John (April 28, 2006). "Biolistic plant transformation". Physiologia Plantarum 79 (1): 206–209. doi:10.1111/j.1399-3054.1990.tb05888.x. Retrieved 16 October 2015.

- Kikkert, Julie; Vidal, Jose; Reisch, Bruce (2005). "Stable transformation of plant cells by particle bombardment/bioistics". Methods of Molecular Biology 286: 61–78. doi:10.1385/1-59259-827-7:061. PMID 15310913. Retrieved 16 October 2015.

- "Expression of an Arabidopsis sodium/proton antiporter gene (AtNHX1) in peanut to improve salt tolerance - Springer". Plant Biotechnology Reports (Link.springer.com) 6: 59–67. 2012-01-01. doi:10.1007/s11816-011-0200-5. Retrieved 2014-02-06.

- Maghari, Behrokh Mohajer, and Ali M. Ardekani. "Genetically Modified Foods And Social Concerns." Avicenna Journal Of Medical Biotechnology 3.3 (2011): 109-117. Academic Search Premier. Web. 7 Nov. 2014.

Fields Related To Agricultural Science

This chapter focuses on those fields whose subject of study is similar to that of agricultural science. These subjects can guide agricultural scientists by offering ideas on moisture content, soil texture, drainage etc. Such ideas can greatly help to improve the growth of modified crops and plants. This chapter is a compilation of the various branches of agricultural science that form an integral part of the broader subject matter.

Agricultural Soil Science

Agricultural soil science is a branch of soil science that deals with the study of edaphic conditions as they relate to the production of food and fiber. In this context, it is also a constituent of the field of agronomy and is thus also described as soil agronomy.

History

Prior to the development of pedology in the 19th century, agricultural soil science (or edaphology) was the only branch of soil science. The bias of early soil science toward viewing soils only in terms of their agricultural potential continues to define the soil science profession in both academic and popular settings as of 2006. (Baveye, 2006)

Current Status

Agricultural soil science follows the holistic method. Soil is investigated in relation to and as integral part of terrestrial ecosystems but is also recognized as a manageable natural resource.

Agricultural soil science studies the chemical, physical, biological, and mineralogical composition of soils as they relate to agriculture. Agricultural soil scientists develop methods that will improve the use of soil and increase the production of food and fiber crops. Emphasis continues to grow on the importance of soil sustainability. Soil degradation such as erosion, compaction, lowered fertility, and contamination continue to be serious concerns. They conduct research in irrigation and drainage, tillage, soil classification, plant nutrition, soil fertility, and other areas.

Although maximizing plant (and thus animal) production is a valid goal, sometimes it may come at high cost which can be readily evident (e.g. massive crop disease stemming from monoculture) or long-term (e.g. impact of chemical fertilizers and pesticides on human health). An agricultural soil scientist may come up with a plan that can maximize production using sustainable methods and solutions, and in order to do that he must look into a number of science fields including agricultural science, physics, chemistry, biology, meteorology and geology.

Soil Variables

Some soil variables of special interest to agricultural soil science are:

- Soil texture or soil composition: Soils are composed of solid particles of various sizes. In decreasing order, these particles are sand, silt and clay. Every soil can be classified according to the relative percentage of sand, silt and clay it contains.

- Aeration and porosity: Atmospheric air contains elements such as oxygen, nitrogen, carbon and others. These elements are prerequisites for life on Earth. Particularly, all cells (including root cells) require oxygen to function and if conditions become anaerobic they fail to respire and metabolize. Aeration in this context refers to the mechanisms by which air is delivered to the soil. In natural ecosystems soil aeration is chiefly accomplished through the vibrant activity of the biota. Humans commonly aerate the soil by tilling and plowing, yet such practice may cause degradation. Porosity refers to the air-holding capacity of the soil.

- Drainage: In soils of bad drainage the water delivered through rain or irrigation may pool and stagnate. As a result, prevail anaerobic conditions and plant roots suffocate. Stagnant water also favors plant-attacking water molds. In soils of excess drainage, on the other hand, plants don't get to absorb adequate water and nutrients are washed from the porous medium to end up in groundwater reserves.

- Water content: Without soil moisture there is no transpiration, no growth and plants wilt. Technically, plant cells loose their pressure. Plants contribute directly to soil moisture. For instance, they create a leafy cover that minimizes the evaporative effects of solar radiation. But even when plants or parts of plants die, the decaying plant matter produces a thick organic cover that protects the soil from evaporation, erosion and compaction.

- Water potential: Water potential describes the tendency of the water to flow from one area of the soil to another. While water delivered to the soil surface normally flows downward due to gravity, at some point it meets increased pressure which causes a reverse upward flow. This effect is known as water suction.

- Horizonation: Typically found in advanced and mature soils, horizonation refers to the creation of soil layers with differing characteristics. It affects almost all soil variables.

- Fertility: A fertile soil is one rich in nutrients and organic matter. Modern agricultural methods have rendered much of the arable land infertile. In such cases, soil can no longer support on its own plants with high nutritional demand and thus needs an external source of nutrients. However, there are cases where human activity is thought to be responsible for transforming rather normal soils into super-fertile ones.

- Biota and soil biota: Organisms interact with the soil and contribute to its quality in innumerable ways. Sometimes the nature of interaction may be unclear, yet a rule is becoming evident: The amount and diversity of the biota is "proportional" to the quality of the soil. Clades of interest include bacteria, fungi, nematodes, annelids and arthropods.

- Soil acidity or soil pH and cation-exchange capacity: Root cells act as hydrogen pumps and the surrounding concentration of hydrogen ions affects their ability to absorb nutrients. pH is a measure of this concentration. Each plant species achieves maximum growth in a particular pH range, yet the vast majority of edible plants can grow in soil pH between 5.0 and 7.5.

Soil Fertility

Agricultural soil scientists study ways to make soils more productive. They classify soils and test them to determine whether they contain nutrients vital to plant growth. Such nutritional substances include compounds of nitrogen, phosphorus, and potassium. If a certain soil is deficient in these substances, fertilizers may provide them. Agricultural soil scientists investigate the movement of nutrients through the soil, and the amount of nutrients absorbed by a plant's roots. Agricultural soil scientists also examine the development of roots and their relation to the soil. Some agricultural soil scientists try to understand the structure and function of soils in relation to soil fertility. They grasp the structure of soil as porous solid. The solid frames of soil consist of mineral derived from the rocks and organic matter originated from the dead bodies of various organisms. The pore space of the soil is essential for the soil to become productive. Small pores serve as water reservoir supplying water to plants and other organisms in the soil during the rain-less period. The water in the small pores of soils is not pure water; they call it soil solution. In soil solution, various plant nutrients derived from minerals and organic matters in the soil are there. This is measured through the cation exchange capacity. Large pores serve as water drainage pipe to allow the excessive water pass through the soil, during the heavy rains. They also serve as air tank to supply oxygen to plant roots and other living beings in the soil. In short, agricultural soil scientists see the soil as a vessel, the most precious one for us, containing all of the substances needed by the plants and other living beings on earth.

Soil Preservation

In addition, agricultural soil scientists develop methods to preserve the agricultural productivity of soil and to decrease the effects on productivity of erosion by wind and water. For example, a technique called contour plowing may be used to prevent soil erosion and conserve rainfall. Researchers in agricultural soil science also seek ways to use the soil more effectively in addressing associated challenges. Such challenges include the beneficial reuse of human and animal wastes using agricultural crops; agricultural soil management aspects of preventing water pollution and the build-up in agricultural soil of chemical pesticides.

Employment of Agricultural Soil Scientists

Most agricultural soil scientists are consultants, researchers, or teachers. Many work in the developed world as farm advisors, agricultural experiment stations, federal, state or local government agencies, industrial firms, or universities. Within the USA they may be trained through the USDA's Cooperative Extension Service offices, although other countries may use universities, research institutes or research agencies. Elsewhere, agricultural soil scientists may serve in international organizations such as the Agency for International Development and the Food and Agriculture Organization of the United Nations.

Quotations

[The key objective of the soil science discipline is that of] finding ways to meet growing human needs for food and fiber while maintaining environmental stability and conserving resources for future generations

—John W. Doran, 2002 SSSA President, 2002

Many people have the vague notion that soil science is merely a phase of agronomy and deals only with practical soil management for field crops. Whether we like it or not this is the image many have of us

—Charles E. Kellog, 1961

Agronomy

Agronomy is the science and technology of producing and using plants for food, fuel, fiber, and land reclamation. Agronomy has come to encompass work in the areas of plant genetics, plant physiology, meteorology, and soil science. It is the application of a combination of sciences like biology, chemistry, economics, ecology, earth science, and genetics. Agronomists of today are involved with many issues, including producing food, creating healthier food, managing the environmental impact of agriculture, and extracting energy from plants. Agronomists often specialise in areas such as crop rotation, irrigation and drainage, plant breeding, plant physiology, soil classification, soil fertility, weed control, and insect and pest control.

Plant Breeding

An agronomist field sampling a trial plot of flax.

This area of agronomy involves selective breeding of plants to produce the best crops under various conditions. Plant breeding has increased crop yields and has improved the nutritional value of numerous crops, including corn, soybeans, and wheat. It has also led to the development of new types of plants. For example, a hybrid grain called triticale was produced by crossbreeding rye and wheat. Triticale contains more usable protein than does either rye or wheat. Agronomy has also been instrumental in fruit and vegetable production research.

Biotechnology

Purdue University agronomy professor George Van Scoyoc explains the difference between forest and prairie soils to soldiers of the Indiana National Guard's Agribusiness Development Team at the Beck Agricultural Center in West Lafayette, Indiana

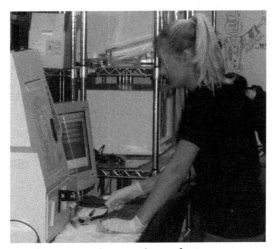

An agronomist mapping a plant genome

Agronomists use biotechnology to extend and expedite the development of desired characteristic. Biotechnology is often a lab activity requiring field testing of the new crop varieties that are developed.

In addition to increasing crop yields agronomic biotechnology is increasingly being applied for novel uses other than food. For example, oilseed is at present used mainly for margarine and other food oils, but it can be modified to produce fatty acids for detergents, substitute fuels and petrochemicals.

Soil Science

Agronomists study sustainable ways to make soils more productive and profitable. They classify soils and analyze them to determine whether they contain nutrients vital to plant growth. Common macronutrients analyzed include compounds of nitrogen, phosphorus, potassium, calcium, magnesium, and sulfur. Soil is also assessed for several micronutrients, like zinc and boron. The percentage of organic matter, soil pH, and nutrient holding capacity (cation exchange capacity) are tested in a regional laboratory. Agronomists will interpret these lab reports and make recommendations to balance soil nutrients for optimal plant growth.

Soil Conservation

In addition, agronomists develop methods to preserve the soil and to decrease the effects of erosion by wind and water. For example, a technique called contour plowing may be used to prevent soil erosion and conserve rainfall. Researchers in agronomy also seek ways to use the soil more effectively in solving other problems. Such problems include the disposal of human and animal manure, water pollution, and pesticide build-up in the soil. Techniques include no-tilling crops, planting of soil-binding grasses along contours on steep slopes, and contour drains of depths up to 1 metre.

Agroecology

Agroecology is the management of agricultural systems with an emphasis on ecological and environmental perspectives. This area is closely associated with work in the areas of sustainable agriculture, organic farming, alternative food systems and the development of alternative cropping systems.

Agronomy Schools

Agronomy programs, offered at colleges, universities, and specialized agricultural schools, often involve classes across a range of departments, including agriculture, biology, chemistry, economics, and physiology. Completing the coursework usually takes from four to twelve years. Many companies will pay agronomists-in-training's college expenses if they agree to work for them when they graduate.

9

Pollination Management: An Integrated Study

Pollination is defined as the natural process of plant reproduction. It is vital to the plant's survival and spread. The advance of technology has meant that it is possible to harness resources and elements that aid in the process of pollination. The chapter strategically encompasses and incorporates the major components and key concepts of pollination management, providing a complete understanding.

Pollination Management

Honey bees are especially well adapted to collecting and moving pollen,thus are the most commonly used crop pollinators. Note the light brown pollen in the pollen basket.

Honey bee on domestic plum blossom

Pollination management is the label for horticultural practices that accomplish or enhance pollination of a crop, to improve yield or quality, by understanding of the particular crop's pollination needs, and by knowledgeable management of pollenizers, pollinators, and pollination conditions.

Placing honey bees for pumpkin pollination
Mohawk Valley, NY

Date pollinator up an 'Abid Rahim' palm tree

Pollinator Decline

With the decline of both wild and domestic pollinator populations, pollination management is becoming an increasingly important part of horticulture. Factors that cause the loss of pollinators include pesticide misuse, unprofitability of beekeeping for honey, rapid transfer of pests and diseases to new areas of the globe, urban/suburban development, changing crop patterns, clearcut logging (particularly when mixed forests are replaced by monoculture pine), clearing of hedgerows and other wild areas, bad diet because of loss of floral biodiversity, and a loss of nectar corridors for migratory pollinators.

Importance

The increasing size of fields and orchards (monoculture) increase the importance of pollination management. Monoculture can cause a brief period when pollinators have more food resources than they can use (but monofloral diet can reduce their immune system) while other periods of the year can bring starvation or pesticide contamination of food sources. Most nectar source and pollen source throughout the growing season to build up their numbers.

Crops that traditionally have had managed pollination include apple, almonds, pears, some plum and cherry varieties, blueberries, cranberries, cucumbers, cantaloupe, watermelon, alfalfa seeds, onion seeds, and many others. Some crops that have traditionally depended entirely on chance pollination by wild pollinators need pollination management nowadays to make a profitable crop.

Some crops, especially when planted in a monoculture situation, require a very high level of pollinators to produce economically viable crops. This may be because of lack of attractiveness of the blossoms, or from trying to pollinate with an alternative when the native pollinator is extinct or rare. These include crops such as alfalfa, cranberries, and kiwifruit. This technique is known as saturation pollination. In many such cases, various native bees are vastly more efficient at pollination (e.g., with blueberries), but the inefficiency of the honey bees is compensated for by using large numbers of hives, the total number of foragers thereby far exceeding the local abundance of native pollinators. In a very few cases, it has been possible to develop commercially viable pollination techniques that use the more efficient pollinators, rather than continued reliance on honey bees, as in the management of the alfalfa leafcutter bee.

Number of hives needed per unit area of crop pollination			
Common name	number of hives per acre	number of hives per hectare	number of bee visits per square meter/minute
Alfalfa	1, (3–5)	2.5, (4.9–12)	
Almonds	2–3	4.9–7.4	
Apples (normal size)	1	2.5	
Apples (semi dwarf)	2	4.9	
Apples (dwarf)	3	7.4	
Apricots	1	2.5	
Blueberries	3–4	7.4–9.9	2.5
Borage	0.6–1.0	1.5–2.5	
Buckwheat	0.5–1	1.2–2.5	
Canola	1	2.5	
Canola (hybrid)	2.0–2.5	4.9–6.2	
Cantaloupes	2–4, (average 2.4)	4.9–9.9, (average 5.9)	
Clovers	1–2	2.5–4.9	
Cranberries	3	7.4	
Cucumbers	1–2, (average 2.1)	2.5–4.9, (average 5.2)	

Number of hives needed per unit area of crop pollination			
Common name	number of hives per acre	number of hives per hectare	number of bee visits per square meter/minute
Ginseng	1	2.5	
Muskmelon	1–3	2.5–7.4	
Nectarines	1	2.5	
Peaches	1	2.5	
Pears	1	2.5	
Plums	1	2.5	
Pumpkins	1	2.5	
Raspberries	0.7–1.3	1.7–3.2	
Squash	1–3	2.5–7.4	
Strawberries	1–3.5	2.5–8.6	
Sunflower	1	2.5	
Trefoil	0.6–1.5	1.5–3.7	
Watermelon	1–3, (average 1.3)	2.5–4.9, (average 3.2)	
Zucchini	1	2.5	

It is estimated that about one hive per acre will sufficiently pollinate watermelons. In the 1950s when the woods were full of wild bee trees, and beehives were normally kept on most South Carolina farms, a farmer who grew ten acres (4 ha) of watermelons would be a large grower and probably had all the pollination needed. But today's grower may grow 200 acres (80 ha), and, if lucky, there might be one bee tree left within range. The only option in the current economy is to bring beehives to the field during blossom time.

Types Of Pollinators

Organisms that are currently being used as pollinators in managed pollination are honey bees, bumblebees, alfalfa leafcutter bees, and orchard mason bees. Other species are expected to be added to this list as this field develops. Humans also can be pollinators, as the gardener who hand pollinates her squash blossoms, or the Middle Eastern farmer, who climbs his date palms to pollinate them.

The Cooperative extension service recommends one honey bee hive per acre (2.5 hives per hectare) for standard watermelon varieties to meet this crop's pollination needs. In the past, when fields were small, pollination was accomplished by a mix of bees kept on farms, bumblebees, carpenter bees, feral honey bees in hollow trees and other insects. Today, with melons planted in large tracts, the grower may no longer have hives on the farm; he may have poisoned many of the pollinators by spraying blooming cotton; he may have logged off the woods, removing hollow trees that provided homes for bees, and pushed out the hedgerows that were home for solitary native bees and other pollinating insects.

Planning for Improved Pollination

Before pollination needs were understood, orchardists often planted entire blocks of apples of a single variety. Because apples are self-sterile, and different members of a single variety are genetic clones (equivalent to a single plant), this is not a good idea. Growers now supply pollenizers, by planting crab apples interspersed in the rows, or by grafting crab apple limbs on some trees. Pollenizers can also be supplied by putting drum bouquets of crab apples or a compatible apple variety in the orchard blocks.

US migratory commercial beekeeper moving spring bees from South Carolina to Maine for blueberry pollination

The field of pollination management cannot be placed wholly within any other field, because it bridges several fields. It draws from horticulture, apiculture, zoology (especially entomology), ecology, and botany.

Improving Pollination with Suboptimal Bee Densities

Growers' demand for beehives far exceeds the available supply. The number of managed beehives in the US has steadily declined from close to 6 million after WWII, to less than 2.5 million today. In contrast, the area dedicated to growing bee-pollinated crops has grown over 300% in the same time period. To make matters worse, in the past five years we have seen a decline in winter managed beehives, which has reached an unprecedented rate near 30%. At present, there is an enormous demand for beehive rentals that cannot always be met. There is a clear need across the agricultural industry for a management tool to draw pollinators into cultivations and encourage them to preferentially visit and pollinate the flowering crop. By attracting pollinators like honeybees and increasing their foraging behavior, particularly in the center of large plots, we can increase grower returns and optimize yield from their plantings.

References

- Frank J. Dainello & Roland Roberts. "Cultural Practices". Texas Vegetable Grower's Handbook. Texas Agricultural Extension Service. Retrieved 9 December 2014.

10

Agriculture Resources

Agricultural resources come to involve agricultural land as well as any tools, techniques and procedures that transform it into productive arable land. Discussed in this chapter are articles dealing with agricultural resources, agricultural machinery and fertilizer.

Agricultural Land

Agricultural land is typically land *devoted to* agriculture, the systematic and controlled use of other forms of life—particularly the rearing of livestock and production of crops—to produce food for humans. It is thus generally synonymous with farmland or cropland.

Photo showing piece of agricultural land irrigated and ploughed for paddy cultivation.

The United Nations Food and Agriculture Organization and others following its definitions, however, also use agricultural land or agricultural area as a term of art, where it means the collection of:

- "arable land" (aka cropland): here redefined to refer to land producing crops requiring annual replanting or fallowland or pasture used for such crops within any five-year period

- "permanent cropland": land producing crops which do not require annual replanting

- permanent pastures: natural or artificial grasslands and shrublands able to be used for grazing livestock

This sense of "agricultural land" thus includes a great deal of land not actively or even presently devoted to agricultural use. The land actually under annually-replanted crops in any given year is instead said to constitute "sown land" or "cropped land". "Permanent cropland" includes forested plantations used to harvest coffee, rubber, or fruit but not tree farms or proper forests used for wood or timber. Land able to be used for farming (traditionally called arable land but here described as "arable land" and "permanent cropland" together) is called **cultivable land**. Farmland, meanwhile, is used variously in reference to all agricultural land, to all cultivable land, or just to the newly restricted sense of "arable land". Depending upon its use of artificial irrigation, the FAO's "agricultural land" may be divided into irrigated and non-irrigated land.

In the context of zoning, agricultural land or agriculturally-zoned land refers to plots that are *permitted* to be used for agricultural activities, without regard to its present use or even suitability. In some areas, agricultural land is protected so that it can be farmed without any threat of development. The Agricultural Land Reserve in British Columbia, for instance, requires approval from its Agricultural Land Commission before its lands can be removed or subdivided.

Area

Under the FAO's definitions above, agricultural land covers 38.4% of the world's land area as of 2011. Permanent pastures are 68.4% of all agricultural land (26.3% of global land area), arable land (row crops) is 28.4% of all agricultural land (10.9% of global land area), and permanent crops (e.g. vineyards and orchards) are 3.1% (1.2% of global land area).

- Total of land used to produce food: 49,116,227 square kilometers or 18,963,881 square miles

- Arable land: 13,963,743 square kilometers or 5,391,431 square miles

- Permanent crops: 1,537,338 square kilometers or 593,570 square miles

- Permanent pastures: 33,585,676 square kilometers or 12,967,502 square miles

Globally, the total amount of agricultural land according to FAO has been in decline since 1998, although it should be noted that this does not account for gross conversion (e.g. land is being extensively cleared for agriculture in some areas, while converted from agriculture to other uses such as urban elsewhere) and more detailed analyses have sometimes found the FAO data to be inaccurate. For example, Lark et al. 2015 found that in the United States cropland increased by 2.98 million acres from 2008-2012 (comprising 7.34 million acres (29,700 km^2) converted to agriculture, and 4.36 million acres (17,600 km^2) converted from agriculture); the FAO shows that during this period U.S. agricultural land declined by 16.56 million acres (67,000 km^2).

Agricultural Land Area ('000 km²)				
	2008	**2009**	**2010**	**2011**
USA	4,044	4,035	4,109	4,113
Germany	169	169	167	167

Helgi Library, World Bank, FAOSTAT

Russia

The cost of Russian farmland is as little as €1,500-2,000/ha (£1,260-1,680/ha). Farmland can be available in France for roughly €10,000/ha, but this is a bargain; for quality soil, realistic prices vary between €50,000-100,000/ha . Farmland has been seen to be available on the Spanish market for as little as €10,000/ha, but this is non-irrigated land.

The average Russian farm measures 150ha. The most prevalent crops in Russia are wheat, barley, corn, rice, sugar beet, soy beans, sunflower, potatoes and vegetables. The Krasnodar region in Russia has 86,000ha of arable land. Russian farmers harvested roughly 85-90 million tonnes of wheat annually in the years around 2010. Russia exported most to Egypt, Turkey and Iran in 2012; China was a significant export market as well. The average yield from the Krasnodar region was between 4 and 5 tonnes per ha, while the Russian average was only 2t/ha. The Basic Element Group, which is a conglomerate owned by Oleg Deripaska, is one of Russia's leading agricultural producers, and owns or manages 109,000ha of Russian farmland, out of 90m actual and 115m total (0.12% actual).

Ukraine

In 2013, Ukraine was ranked third in corn production and sixth in wheat production. It was the main supplier of corn, wheat, and rape to Europe, although it is unclear whether the internal supply from countries like France were accounted in this calculation. Ukrainian farmers achieve 60% of the output per unit area of their North American competitors. UkrLandFarming PLC produces from 1.6m acres corn wheat barley sugar beet and sunflowers. Until 2014, the chief Ukrainian export terminal was the Crimean port of Sebastopol.

Agricultural Machinery

A German combine harvester by Claas

Agricultural machinery is machinery used in the operation of an agricultural area or farm.

A British crop sprayer by Lite-Trac

History of the Machines

The Industrial Revolution

With the coming of the Industrial Revolution and the development of more complicated machines, farming methods took a great leap forward. Instead of harvesting grain by hand with a sharp blade, wheeled machines cut a continuous swath. Instead of threshing the grain by beating it with sticks, threshing machines separated the seeds from the heads and stalks. The first tractors appeared in the late 19th century.

Steam Power

Power for agricultural machinery was originally supplied by ox or other domesticated animals. With the invention of steam power came the portable engine, and later the traction engine, a multipurpose, mobile energy source that was the ground-crawling cousin to the steam locomotive. Agricultural steam engines took over the heavy pulling work of oxen, and were also equipped with a pulley that could power stationary machines via the use of a long belt. The steam-powered machines were low-powered by today's standards but, because of their size and their low gear ratios, they could provide a large drawbar pull. Their slow speed led farmers to comment that tractors had two speeds: "slow, and damn slow."

Internal Combustion Engines

The internal combustion engine; first the petrol engine, and later diesel engines; became the main source of power for the next generation of tractors. These engines also contributed to the development of the self-propelled, combined harvester and thresher, or combine harvester (also shortened to 'combine'). Instead of cutting the grain stalks and transporting them to a stationary threshing machine, these combines cut, threshed, and separated the grain while moving continuously through the field.

Types

A John Deere cotton harvester at work in a cotton field.

From left to right: John Deere 7800 tractor with Houle slurry trailer, Case IH combine harvester, New Holland FX 25 forage harvester with corn head.

A New Holland TR85 combine harvester

Combines might have taken the harvesting job away from tractors, but tractors still do the majority of work on a modern farm. They are used to push implements—machines that till the ground, plant seed, and perform other tasks.

Tillage implements prepare the soil for planting by loosening the soil and killing weeds or competing plants. The best-known is the plow, the ancient implement that was upgraded in 1838 by John Deere. Plows are now used less frequently in the U.S. than formerly, with offset disks used instead to turn over the soil, and chisels used to gain the depth needed to retain moisture.

The most common type of seeder is called a planter, and spaces seeds out equally in long rows, which are usually two to three feet apart. Some crops are planted by drills, which put out much more seed in rows less than a foot apart, blanketing the field with crops. Transplanters automate the task of transplanting seedlings to the field. With the widespread use of plastic mulch, plastic mulch layers, transplanters, and seeders lay down long rows of plastic, and plant through them automatically.

After planting, other implements can be used to cultivate weeds from between rows, or to spread fertilizer and pesticides. Hay balers can be used to tightly package grass or alfalfa into a storable form for the winter months.

Modern irrigation relies on machinery. Engines, pumps and other specialized gear provide water quickly and in high volumes to large areas of land. Similar types of equipment can be used to deliver fertilizers and pesticides.

Besides the tractor, other vehicles have been adapted for use in farming, including trucks, airplanes, and helicopters, such as for transporting crops and making equipment mobile, to aerial spraying and livestock herd management.

New Technology and the Future

The basic technology of agricultural machines has changed little in the last century. Though modern harvesters and planters may do a better job or be slightly tweaked from their predecessors, the US$250,000 combine of today still cuts, threshes, and separates grain in the same way it has always been done. However, technology is changing the way that humans operate the machines, as computer monitoring systems, GPS locators, and self-steer programs allow the most advanced tractors and implements to be more precise and less wasteful in the use of fuel, seed, or fertilizer. In the foreseeable future, there may be mass production of driverless tractors, which use GPS maps and electronic sensors.

Open Source Agricultural Equipment

Many farmers are upset by their inability to fix the new types of high-tech farm equipment. This is due mostly to companies using intellectual property law to prevent farmers from having the legal right to fix their equipment (or gain access to the information to allow them to do it). This has encouraged groups such as Open Source Ecology and Farm Hack to begin to make open source agricultural machinery. In addition on a smaller scale FarmBot and the RepRap open source 3D printer community has begun to make open-source farm tools available of increasing levels of sophistication. In October 2015 an exemption was added to the DMCA to allow inspection and modification of the software in cars and other vehicles including agricultural machinery.

Fertilizer

A fertilizer (American English) or fertiliser (British English) is any material of natural or synthetic origin (other than liming materials) that is applied to soils or to plant tissues (usually leaves) to supply one or more plant nutrients essential to the growth of plants.

A large, modern fertilizer spreader

A Lite-Trac Agri-Spread lime and fertilizer spreader at an agricultural show

Mechanism

Six tomato plants grown with and without nitrate fertilizer on nutrient-poor sand/clay soil. One of the plants in the nutrient-poor soil has died.

Fertilizers enhance the growth of plants. This goal is met in two ways, the traditional one being additives that provide nutrients. The second mode by some fertilizers act is to enhance the effectiveness of the soil by modifying its water retention and aeration. This article, like many on fertilizers, emphasises the nutritional aspect. Fertilizers typically provide, in varying proportions:

- three main macronutrients:

  o Nitrogen (N): leaf growth;

  o Phosphorus (P): Development of roots, flowers, seeds, fruit;

  o Potassium (K): Strong stem growth, movement of water in plants, promotion of flowering and fruiting;

- three secondary macronutrients: calcium (Ca), magnesium (Mg), and sulphur (S);

- micronutrients: copper (Cu), iron (Fe), manganese (Mn), molybdenum (Mo), zinc (Zn), boron (B), and of occasional significance there are silicon (Si), cobalt (Co), and vanadium (V) plus rare mineral catalysts.

The nutrients required for healthy plant life are classified according to the elements, but the elements are not used as fertilizers. Instead compounds containing these elements are the basis of fertilisers. The macronutrients are consumed in larger quantities and are present in plant tissue in quantities from 0.15% to 6.0% on a dry matter (DM) (0% moisture) basis. Plants are made up of four main elements: hydrogen, oxygen, carbon, and nitrogen. Carbon, hydrogen and oxygen are widely available as water and carbon dioxide. Although nitrogen makes up most of the atmosphere, it is in a form that is unavailable to plants. Nitrogen is the most important fertilizer since nitrogen is present in proteins, DNA and other components (e.g., chlorophyll). To be nutritious to plants, nitrogen must be made available in a "fixed" form. Only some bacteria and their host plants (notably legumes) can fix atmospheric nitrogen (N_2) by converting it to ammonia. Phosphate is required for the production of DNA and ATP, the main energy carrier in cells, as well as certain lipids.

Micronutrients are consumed in smaller quantities and are present in plant tissue on the order of parts-per-million (ppm), ranging from 0.15 to 400 ppm DM, or less than 0.04% DM. These elements are often present at the active sites of enzymes that carry out the plant's metabolism. Because these elements enable catalysts (enzymes) their impact far exceeds their weight percentage.

Classification

Fertilizers are classified in several ways. They are classified according to whether they provide a single nutrient (say, N, P, or K), in which case they are classified as "straight fertilizers." "Multinutrient fertilizers" (or "complex fertilizers") provide two or more nutrients, for example N and P. Fertilizers are also sometimes classified as inorganic (the topic of most of this article) versus organic. Inorganic fertilizers exclude carbon-containing materials except ureas. Organic fertilizers are usually (recycled) plant- or animal-derived matter. Inorganic are sometimes called synthetic fertilizers since various chemical treatments are required for their manufacture.

Single Nutrient ("Straight") Fertilizers

The main nitrogen-based straight fertilizer is ammonia or its solutions. Ammonium nitrate (NH_4NO_3) is also widely used. About 15M tons were produced in 1981. Urea is another popular source of nitrogen, having the advantage that it is a solid and non-explosive, unlike ammonia and ammonium nitrate, respectively. A few percent of the nitrogen fertilizer market (4% in 2007) has been met by calcium ammonium nitrate ($Ca(NO_3)_2 \cdot NH_4NO_3 \cdot 10H_2O$).

The main straight phosphate fertilizers are the superphosphates. "Single superphosphate" (SSP) consists of 14–18% P_2O_5, again in the form of $Ca(H_2PO_4)_2$, but also phosphogypsum ($CaSO_4 \cdot 2 H_2O$). Triple superphosphate (TSP) typically consists of 44-48% of P_2O_5 and no gypsum. A mixture of single superphosphate and triple superphosphate is called double superphosphate. More than 90% of a typical superphosphate fertilizer is water-soluble.

Multinutrient Fertilizers

These fertilizers are the most common. They consist of two or more nutrient components.

Binary (NP, NK, PK) Fertilizers

Major two-component fertilizers provide both nitrogen and phosphorus to the plants. These are called NP fertilizers. The main NP fertilizers are monoammonium phosphate (MAP) and diammonium phosphate (DAP). The active ingredient in MAP is $NH_4H_2PO_4$. The active ingredient in DAP is $(NH_4)_2HPO_4$. About 85% of MAP and DAP fertilizers are soluble in water.

Npk Fertilizers

NPK fertilizers are three-component fertilizers providing nitrogen, phosphorus, and potassium.

NPK rating is a rating system describing the amount of nitrogen, phosphorus, and potassium in a fertilizer. NPK ratings consist of three numbers separated by dashes (e.g., 10-10-10 or 16-4-8) describing the chemical content of fertilizers. The first number represents the percentage of nitrogen in the product; the second number, P_2O_5; the third, K_2O. Fertilizers do not actually contain P_2O_5 or K_2O, but the system is a conventional shorthand for the amount of the phosphorus (P) or potassium (K) in a fertilizer. A 50-pound (23 kg) bag of fertilizer labeled 16-4-8 contains 8 lb (3.6 kg) of nitrogen (16% of the 50 pounds), an amount of phosphorus equivalent to that in 2 pounds of P_2O_5 (4% of 50 pounds), and 4 pounds of potassium K_2O (8% of 50 pounds). Most fertilizers are labeled according to this N-P-K convention, although Australian convention, following an N-P-K-S system, adds a fourth number for sulfur.

Micronutrients

The main micronutrients are molybdenum, zinc, and copper. these elements are provided as water-soluble salts. Iron presents special problems because it converts to insoluble (bio-unavailable) compounds at moderate soil pH and phosphate concentrations. For this reason, iron is often administered as a chelate complex, e.g., the EDTA derivative. The micronutrient needs depend on the plant. For example, sugar beets appear to require boron, and legumes require cobalt.

Production

Nitrogen Fertilizers

Top users of nitrogen-based fertilizer		
Country	Total N use (Mt pa)	Amt. used for feed/pasture (Mt pa)
China	18.7	3.0
India	11.9	N/A
U.S.	9.1	4.7
France	2.5	1.3

Germany	2.0	1.2
Brazil	1.7	0.7
Canada	1.6	0.9
Turkey	1.5	0.3
UK	1.3	0.9
Mexico	1.3	0.3
Spain	1.2	0.5
Argentina	0.4	0.1

Nitrogen fertilizers are made from ammonia (NH_3), which is sometimes injected into the ground directly. The ammonia is produced by the Haber-Bosch process. In this energy-intensive process, natural gas (CH_4) supplies the hydrogen, and the nitrogen (N_2) is derived from the air. This ammonia is used as a feedstock for all other nitrogen fertilizers, such as anhydrous ammonium nitrate (NH_4NO_3) and urea ($CO(NH_2)_2$).

Deposits of sodium nitrate ($NaNO_3$) (Chilean saltpeter) are also found in the Atacama desert in Chile and was one of the original (1830) nitrogen-rich fertilizers used. It is still mined for fertilizer.

There has been technical work investigating on-site (on-farm) synthesis of nitrate fertilizer using solar photovoltaic power, which would enable farmers more control in soil fertility, while using far less surface area than conventional organic farming for nitrogen fertilizer.

Phosphate Fertilizers

All phosphate fertilizers are obtained by extraction from minerals containing the anion PO_4^{3-}. In rare cases, fields are treated with the crushed mineral, but most often more soluble salts are produced by chemical treatment of phosphate minerals. The most popular phosphate-containing minerals are referred to collectively as phosphate rock. The main minerals are fluorapatite $Ca_5(PO_4)_3F$ (CFA) and hydroxyapatite $Ca_5(PO_4)_3OH$. These minerals are converted to water-soluble phosphate salts by treatment with sulfuric or phosphoric acids. The large production of sulfuric acid as an industrial chemical is primarily due to its use as cheap acid in processing phosphate rock into phosphate fertilizer. The global primary uses for both sulfur and phosphorus compounds relate to this basic process.

In the nitrophosphate process or Odda process (invented in 1927), phosphate rock with up to a 20% phosphorus (P) content is dissolved with nitric acid (HNO_3) to produce a mixture of phosphoric acid (H_3PO_4) and calcium nitrate ($Ca(NO_3)_2$). This mixture can be combined with a potassium fertilizer to produce a *compound fertilizer* with the three macronutrients N, P and K in easily dissolved form.

Potassium Fertilizers

Potash is a mixture of potassium minerals used to make potassium (chemical symbol: K) fertilizers. Potash is soluble in water, so the main effort in producing this nutrient from the ore involves some purification steps; e.g., to remove sodium chloride (NaCl) (common salt). Sometimes potash is referred to as K_2O, as a matter of convenience to those describing the potassium content. In fact potash fertilizers are usually potassium chloride, potassium sulfate, potassium carbonate, or potassium nitrate.

Compound Fertilizers

Compound fertilizers, which contain N, P, and K, can often be produced by mixing straight fertilizers. In some cases, chemical reactions occur between the two or more components. For example, monoammonium and diammonium phosphates, which provide plants with both N and P, are produced by neutralizing phosphoric acid (from phosphate rock) and ammonia :

$$NH_3 + H_3PO_4 \rightarrow (NH_4)H_2PO_4$$

$$2\ NH_3 + H_3PO_4 \rightarrow (NH_4)_2HPO_4$$

Organic Fertilizers

Compost bin for small-scale production of organic fertilizer

The main "organic fertilizers" are peat, animal wastes, plant wastes from agriculture, and treated sewage sludge (biosolids). In terms of volume, peat is the most widely used organic fertilizer. This immature form of coal confers no nutritional value to the plants, but improves the soil by aeration and absorbing water. Animal sources include the products of the slaughter of animals. Bloodmeal, bone meal, hides, hoofs, and horns are typical components. Organic fertilizer usually contain fewer nutrients, but offer other advantages as well as being appealing to those who are trying to practice "environmentally friendly" farming.

A large commercial compost operation

Other Elements: Calcium, Magnesium, and Sulfur

Calcium is supplied as superphosphate or calcium ammonium nitrate solutions.

Application

Fertilizers are commonly used for growing all crops, with application rates depending on the soil fertility, usually as measured by a soil test and according to the particular crop. Legumes, for example, fix nitrogen from the atmosphere and generally do not require nitrogen fertilizer.

Liquid vs Solid

Fertilizers are applied to crops both as solids and as liquid. About 90% of fertilizers are applied as solids. Solid fertilizer is typically granulated or powdered. Often solids are available as prills, a solid globule. Liquid fertilizers comprise anhydrous ammonia, aqueous solutions of ammonia, aqueous solutions of ammonium nitrate or urea. These concentrated products may be diluted with water to form a concentrated liquid fertilizer (e.g., UAN). Advantages of liquid fertilizer are its more rapid effect and easier coverage. The addition of fertilizer to irrigation water is called "fertigation".

Slow- and Controlled-Release Fertilizers

Slow- and controlled-release involve only 0.15% (562,000 tons) of the fertilizer market (1995). Their utility stems from the fact that fertilizers are subject to antagonistic processes. In addition to their providing the nutrition to plants, excess fertilizers can be poisonous to the same plant. Competitive with the uptake by plants is the degradation or loss of the fertilizer. Microbes degrade many fertilizers, e.g., by immobilization or oxidation. Furthermore, fertilizers are lost by evaporation or leaching. Most slow-release fertilizers are derivatives of urea, a straight fertilizer providing nitrogen. Isobutylidenediurea ("IBDU") and urea-formaldehyde slowly convert in the soil to free urea, which is rapidly uptaken by plants. IBDU is a single compound with the formula $(CH_3)_2CH-CH(NHC(O)NH_2)_2$ whereas the urea-formaldehydes consist of mixtures of the approximate formula $(HOCH_2NHC(O)NH)_nCH_2$.

Besides being more efficient in the utilization of the applied nutrients, slow-release technologies also reduce the impact on the environment and the contamination of the subsurface water. Slow-release fertilizers (various forms including fertilizer spikes, tabs, etc.) which reduce the problem of "burning" the plants due to excess nitrogen. Polymer coating of fertilizer ingredients gives tablets and spikes a 'true time-release' or 'staged nutrient release' (SNR) of fertilizer nutrients.

Controlled release fertilizers are traditional fertilizers encapsulated in a shell that degrades at a specified rate. Sulfur is a typical encapsulation material. Other coated products use thermoplastics (and sometimes ethylene-vinyl acetate and surfactants, etc.) to produce diffusion-controlled release of urea or other fertilizers. "Reactive Layer Coating" can produce thinner, hence cheaper, membrane coatings by applying reactive monomers simultaneously to the soluble particles. "Multicote" is a process applying layers of low-cost fatty acid salts with a paraffin topcoat.

Foliar Application

Foliar fertilizers are applied directly to leaves. The method is almost invariably used to apply water-soluble straight nitrogen fertilizers and used especially for high value crops such as fruits.

Fertilizer burn

Chemicals that Affect Nitrogen Uptake

Various chemicals are used to enhance the efficiency of nitrogen-based fertilizers. In this way farmers can limit the polluting effects of nitrogen run-off. Nitrification inhibitors (also known as nitrogen stabilizers) suppress the conversion of ammonia into nitrate, an anion that is more prone to leaching. 1-Carbamoyl-3-methylpyrazole (CMP), dicyandiamide, and nitrapyrin (2-chloro-6-trichloromethylpyridine) are popular. Urease inhibitors are used to slow the hydrolytic conversion of urea into ammonia, which is prone to evaporation as well as nitrification. The conversion of urea to ammonia catalyzed by enzymes called ureases. A popular inhibitor of ureases is N-(n-butyl) thiophosphoric triamide (NBPT).

Overfertilization

Careful fertilization technologies are important because excess nutrients can be as detrimental. Fertilizer burn can occur when too much fertilizer is applied, resulting in drying out of the leaves and damage or even death of the plant.Fertilizers vary in their tendency to burn roughly in accordance with their salt index.

Statistics

Conservative estimates report 30 to 50% of crop yields are attributed to natural or synthetic commercial fertilizer. Global market value is likely to rise to more than US$185 billion until 2019. The European fertilizer market will grow to earn revenues of approx. €15.3 billion in 2018.

The map displays the statistics of fertilizer consumption in western and central European counties from data published by The World Bank for 2012.

Data on the fertilizer consumption per hectare arable land in 2012 are published by The World Bank. For the diagram below values of the European Union (EU) countries have been extracted and are presented as kilograms per hectare (pounds per acre). The total consumption of fertilizer in the EU is 15.9 million tons for 105 million hectare arable land area (or 107 million hectare arable land according to another estimate). This figure equates to 151 kg of fertilizers consumed per ha arable land on average for the EU countries. Interestingly, mainly in those countries where fertilizers are consumed a lot also plant growth product are sold more than in others.

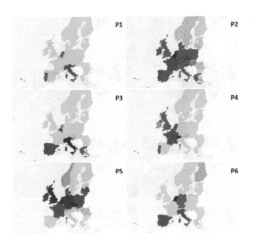

Pesticide categories, EUROSTAT. P5= Plant growth regulators. The red/green scale represents high/low pesticide sales per arable land.

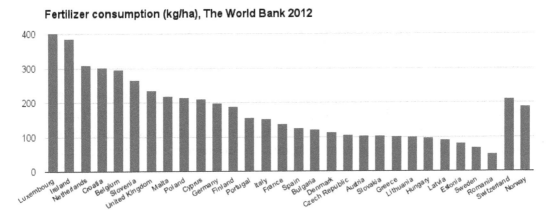

Fertilizer consumption (kg/ha), The World Bank 2012

Environmental Effects

Runoff of soil and fertilizer during a rain storm

An algal bloom caused by eutrophication

Water

Agricultural run-off is a major contributor to the eutrophication of fresh water bodies. For example, in the US, about half of all the lakes are eutrophic. The main contributor to eutrophication is phosphate, which is normally a limiting nutrient; high concentrations promote the growth of cyanobacteria and algae, the demise of which consumes oxygen. Cyanobacteria blooms ('algal blooms') can also produce harmful toxins that can accumulate in the food chain, and can be harmful to humans.

The nitrogen-rich compounds found in fertilizer runoff are the primary cause of serious oxygen depletion in many parts of oceans, especially in coastal zones, lakes and rivers. The resulting lack of dissolved oxygen greatly reduces the ability of these areas to sustain oceanic fauna. The number of oceanic dead zones near inhabited coastlines are increasing. As of 2006, the application of nitrogen fertilizer is being increasingly controlled in northwestern Europe and the United States. If eutrophication *can* be reversed, it may take decades before the accumulated nitrates in groundwater can be broken down by natural processes.

Nitrate Pollution

Only a fraction of the nitrogen-based fertilizers is converted to produce and other plant matter. The remainder accumulates in the soil or lost as run-off. High application rates of nitrogen-containing fertilizers combined with the high water-solubility of nitrate leads to increased runoff into surface water as well as leaching into groundwater, thereby causing groundwater pollution. The excessive use of nitrogen-containing fertilizers (be they synthetic or natural) is particularly damaging, as much of the nitrogen that is not taken up by plants is transformed into nitrate which is easily leached.

Nitrate levels above 10 mg/L (10 ppm) in groundwater can cause 'blue baby syndrome' (acquired methemoglobinemia). The nutrients, especially nitrates, in fertilizers can cause problems for natural habitats and for human health if they are washed off soil into watercourses or leached through soil into groundwater.

Soil

Acidification

Nitrogen-containing fertilizers can cause soil acidification when added. This may lead to decreases in nutrient availability which may be offset by liming.

Accumulation of Toxic Elements

Cadmium

The concentration of cadmium in phosphorus-containing fertilizers varies considerably and can be problematic. For example, mono-ammonium phosphate fertilizer may have a cadmium content of as low as 0.14 mg/kg or as high as 50.9 mg/kg. This is because the phosphate rock used in their manufacture can contain as much as 188 mg/kg cadmium (examples are deposits on Nauru and the Christmas islands). Continuous use of high-cadmium fertilizer can contaminate soil (as shown in New Zealand) and plants. Limits to the cadmium content of phosphate fertilizers has been considered by the European Commission. Producers of phosphorus-containing fertilizers now select phosphate rock based on the cadmium content.

Fluoride

Phosphate rocks contain high levels of fluoride. Consequently, the widespread use of phosphate fertilizers has increased soil fluoride concentrations. It has been found that food contamination from fertilizer is of little concern as plants accumulate little fluoride from the soil; of greater concern is the possibility of fluoride toxicity to livestock that ingest contaminated soils. Also of possible concern are the effects of fluoride on soil microorganisms.

Radioactive Elements

The radioactive content of the fertilizers varies considerably and depends both on their concentrations in the parent mineral and on the fertilizer production process. Uranium-238 concentrations range can range from 7 to 100 pCi/g in phosphate rock and from 1 to 67 pCi/g in phosphate fertilizers. Where high annual rates of phosphorus fertilizer are used, this can result in uranium-238 concentrations in soils and drainage waters that are several times greater than are normally present. However, the impact of these increases on the risk to human health from radinuclide contamination of foods is very small (less than 0.05 mSv/y).

Other Metals

Steel industry wastes, recycled into fertilizers for their high levels of zinc (essential to plant growth), wastes can include the following toxic metals: lead arsenic, cadmium, chromium, and nickel. The most common toxic elements in this type of fertilizer are mercury, lead, and arsenic. These potentially harmful impurities can be removed; however, this significantly increases cost. Highly pure fertilizers are widely available and perhaps best known as the highly water-soluble fertilizers containing blue dyes used around households, such as Miracle-Gro. These highly water-soluble fertilizers are used in the plant nursery business and are available in larger packages at significantly less cost than retail quantities. There are also some inexpensive retail granular garden fertilizers made with high purity ingredients.

Trace Mineral Depletion

Attention has been addressed to the decreasing concentrations of elements such as iron, zinc, copper and magnesium in many foods over the last 50–60 years. Intensive farming practices, including the use of synthetic fertilizers are frequently suggested as reasons for these declines and organic farming is often suggested as a solution. Although improved crop yields resulting from NPK fertilizers are known to dilute the concentrations of other nutrients in plants, much of the measured decline can be attributed to the use of progressively higher-yielding crop varieties which produce foods with lower mineral concentrations than their less productive ancestors. It is, therefore, unlikely that organic farming or reduced use of fertilizers will solve the problem; foods with high nutrient density are posited to be achieved using older, lower-yielding varieties or the development of new high-yield, nutrient-dense varieties.

Fertilizers are, in fact, more likely to solve trace mineral deficiency problems than cause them: In Western Australia deficiencies of zinc, copper, manganese, iron and molybdenum were identified as limiting the growth of broad-acre crops and pastures in the 1940s and 1950s. Soils in Western Australia are very old, highly weathered and deficient in many of the major nutrients and trace elements. Since this time these trace elements are routinely added to fertilizers used in agriculture in this state. Many other soils around the world are deficient in zinc, leading to deficiency in both plants and humans, and zinc fertilizers are widely used to solve this problem.

Changes in Soil Biology

High levels of fertilizer may cause the breakdown of the symbiotic relationships between plant roots and mycorrhizal fungi.

Energy Consumption and Sustainability

In the USA in 2004, 317 billion cubic feet of natural gas were consumed in the industrial production of ammonia, less than 1.5% of total U.S. annual consumption of natural gas. A 2002 report suggested that the production of ammonia consumes about 5% of global natural gas consumption, which is somewhat under 2% of world energy production.

Ammonia is produced from natural gas and air. The cost of natural gas makes up about 90% of the cost of producing ammonia. The increase in price of natural gases over the past decade, along with other factors such as increasing demand, have contributed to an increase in fertilizer price.

Contribution to Climate Change

The greenhouse gases carbon dioxide, methane and nitrous oxide are produced during the manufacture of nitrogen fertilizer. The effects can be combined into an equivalent amount of carbon dioxide. The amount varies according to the efficiency of the process. The figure for the United Kingdom is over 2 kilogrammes of carbon dioxide equivalent for each kilogramme of ammonium nitrate. Nitrogen fertilizer can be converted by soil bacteria to nitrous oxide, a greenhouse gas.

Atmosphere

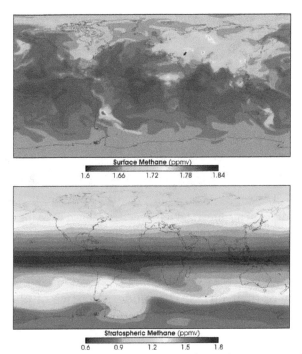

Global methane concentrations (surface and atmospheric) for 2005; note distinct plumes

Through the increasing use of nitrogen fertilizer, which was used at a rate of about 110 million tons (of N) per year in 2012, adding to the already existing amount of reactive nitrogen, nitrous oxide (N_2O) has become the third most important greenhouse gas after carbon dioxide and methane. It has a global warming potential 296 times larger than an equal mass of carbon dioxide and it also contributes to stratospheric ozone depletion. By changing processes and procedures, it is possible to mitigate some, but not all, of these effects on anthropogenic climate change.

Methane emissions from crop fields (notably rice paddy fields) are increased by the application of ammonium-based fertilizers. These emissions contribute to global climate change as methane is a potent greenhouse gas.

Regulation

In Europe problems with high nitrate concentrations in run-off are being addressed by the European Union's Nitrates Directive. Within Britain, farmers are encouraged to manage their land more sustainably in 'catchment-sensitive farming'. In the US, high concentrations of nitrate and phosphorus in runoff and drainage water are classified as non-point source pollutants due to their diffuse origin; this pollution is regulated at state level. Oregon and Washington, both in the United States, have fertilizer registration programs with on-line databases listing chemical analyses of fertilizers.

History

Founded in 1812, Mirat, producer of manures and fertilizers, is claimed to be the oldest industrial business in Salamanca (Spain).

Management of soil fertility has been the preoccupation of farmers for thousands of years. Egyptians, Romans, Babylonians, and early Germans all are recorded as using minerals and or manure to enhance the productivity of their farms. The modern science of plant nutrition started in the 19th century and the work of German chemist Justus von Liebig, among others. John Bennet Lawes, an English entrepreneur, began to experiment on the effects of various manures on plants growing in pots in 1837, and a year or two later the experiments were extended to crops in the field. One immediate consequence was that in 1842 he patented a manure formed by treating phosphates with sulphuric acid, and thus was the first to create the artificial manure industry. In the succeeding year he enlisted the services of Joseph Henry Gilbert, with whom he carried on for more than half a century on experiments in raising crops at the Institute of Arable Crops Research.

The Birkeland–Eyde process was one of the competing industrial processes in the beginning of nitrogen based fertilizer production. This process was used to fix atmospheric nitrogen (N_2) into nitric acid (HNO_3), one of several chemical processes generally referred to as nitrogen fixation. The resultant nitric acid was then used as a source of nitrate (NO_3^-). A factory based on the process was built in Rjukan and Notodden in Norway, combined with the building of large hydroelectric power facilities.

The 1910s and 1920s witness the rise of the Haber process and the Ostwald process. The Haber process produces ammonia (NH_3) from methane (CH_4) gas and molecular nitrogen (N_2). The ammonia from the Haber process is then converted into nitric acid (HNO_3) in the Ostwald process. The development of synthetic fertilizer has significantly supported global population growth — it has been estimated that almost half the people on the Earth are currently fed as a result of synthetic nitrogen fertilizer use.

The use of commercial fertilizers has increased steadily in the last 50 years, rising almost 20-fold to the current rate of 100 million tonnes of nitrogen per year. Without commercial fertilizers it is estimated that about one-third of the food produced now could not be produced. The use of phosphate fertilizers has also increased from 9 million tonnes per year in 1960 to 40 million tonnes per year in 2000. A maize crop yielding 6–9 tonnes of grain per hectare (2.5 acres) requires 31–50 kilograms (68–110 lb) of phosphate fertilizer to be applied; soybean crops require about half, as 20–25 kg per hectare. Yara International is the world's largest producer of nitrogen-based fertilizers.

Controlled-nitrogen-release technologies based on polymers derived from combining urea and formaldehyde were first produced in 1936 and commercialized in 1955. The early product had 60 percent of the total nitrogen cold-water-insoluble, and the unreacted (quick-release) less than 15%. Methylene ureas were commercialized in the 1960s and 1970s, having 25% and 60% of the nitrogen as cold-water-insoluble, and unreacted urea nitrogen in the range of 15% to 30%.

In the 1960s, the Tennessee Valley Authority National Fertilizer Development Center began developing sulfur-coated urea; sulfur was used as the principal coating material because of its low cost and its value as a secondary nutrient. Usually there is another wax or polymer which seals the sulfur; the slow-release properties depend on the degradation of the secondary sealant by soil microbes as well as mechanical imperfections (cracks, etc.) in the sulfur. They typically provide 6 to 16 weeks of delayed release in turf applications. When a hard polymer is used as the secondary coating, the properties are a cross between diffusion-controlled particles and traditional sulfur-coated.

References

- Dittmar, Heinrich; Drach, Manfred; Vosskamp, Ralf; Trenkel, Martin E.; Gutser, Reinhold; Steffens, Günter (2009). "Fertilizers, 2. Types". Ullmann's Encyclopedia of Industrial Chemistry. doi:10.1002/14356007.n10_n01. ISBN 3527306730.

- H.A. Mills; J.B. Jones Jr. (1996). Plant Analysis Handbook II: A practical Sampling, Preparation, Analysis, and Interpretation Guide. ISBN 1-878148-05-2.

- Moore, Geoff (2001). Soilguide - A handbook for understanding and managing agricultural soils (PDF). Perth, Western Australia: Agriculture Western Australia. pp. 161–207. ISBN 0 7307 0057 7.

- Appl, Max (2000). Ullmann's Encyclopedia of Industrial Chemistry, Volume 3. Weinheim, Germany: Wiley-VCH Verlag GmbH & Co. KGaA. pp. 139–225. doi:10.1002/14356007.002_011. ISBN 9783527306732.

- G. J. Leigh (2004). The world's greatest fix: a history of nitrogen and agriculture. Oxford University Press US. pp. 134–139. ISBN 0-19-516582-9.

- Trevor Illtyd Williams; Thomas Kingston Derry (1982). A short history of twentieth-century technology c. 1900-c. 1950. Oxford University Press. pp. 134–135. ISBN 0-19-858159-9.

- Aaron John Ihde (1984). The development of modern chemistry. Courier Dover Publications. p. 678. ISBN 0-486-64235-6.

- FAO (2012). Current world fertilizer trends and outlook to 2016 (PDF). Rome: Food and Agriculture Organization of the United Nations. p. 13. Retrieved 3 July 2014.

- Lugon-Moulin, N.; Ryan, L.; Donini, P.; Rossi, L. (2006). "Cadmium content of phosphate fertilizers used for tobacco production" (PDF). Agron. Sustain. Dev. 26: 151–155. doi:10.1051/agro:2006010. Retrieved 27 June 2014.

- Zapata, F.; Roy, R.N. (2004). "Use of Phosphate Rocks for Sustainable Agriculture: Secondary nutrients, micronutrients, liming effect and hazardous elements associated with phosphate rock use". www.fao.org. FAO. Retrieved 27 June 2014.

- Hanlon, E. A. (2012). "Naturally Occurring Radionuclides in Agricultural Products". edis.ifas.ufl.edu. University of Florida. Retrieved 17 July 2014.

- Zapata, F; Roy, RN (2004). Use of phosphate rocks for sustainable agriculture (PDF). Rome: FAO. p. 82. Retrieved 16 July 2014.

- Roy, R. N.; Misra, R. V.; Montanez, A. (2002). "Decreasing reliance on mineral nitrogen-yet more food" (PDF). AMBIO: A Journal of the Human Environment. 31 (2): 177–183. doi:10.1579/0044-7447-31.2.177. Retrieved 3 July 2014.

- "Label Requirements of specialty and other bagged fertilizers". Michigan Department of Agriculture and Rural Development. Retrieved 14 March 2013.

- Schmidt, JR; Shaskus, M; Estenik, JF; Oesch, C; Khidekel, R; Boyer, GL (2013). "Variations in the microcystin content of different fish species collected from a eutrophic lake". Toxins (Basel). 5: 992–1009. doi:10.3390/toxins5050992. PMC 3709275. PMID 23676698.

- Banger, K.; Tian, H.; Lu, C. (2012). "Do nitrogen fertilizers stimulate or inhibit methane emissions from rice fields?". Global Change Biology. 18 (10): 3259–3267. doi:10.1111/j.1365-2486.2012.02762.x.

Evolution of Agriculture

Agricultural science has coherently recorded and applied the latest advancements that have taken place in the fields of agriculture, cultivation and irrigation. This chapter touches upon the major developments that have taken place in agricultural science in the last century-and-a-half.

History of Agricultural Science

History of agricultural science looks at the scientific advancement of techniques and understanding of agriculture. Scientific study of fertilizer was advanced significantly in 1840 with the publication *Die organische Chemie in ihrer Anwendung auf Agrikulturchemie und Physiologie* (Organic Chemistry in Its Applications to Agriculture and Physiology) by Justus von Liebig. One of Liebig's advances in agricultural science was the discovery of nitrogen as an essential plant nutrient.

Fertilizer

Johann Friedrich Mayer was the first scientist to publish experiments on the use of gypsum as a fertilizer, but the mechanism that made it function as a fertilizer was contested by his contemporaries.

In 1840, Liebig published what is now known as Liebig's law of the minimum. Liebig's law states that growth is not controlled by the total amount of resources available, but by the limiting factor.

In the United States, a scientific revolution in agriculture began with the Hatch Act of 1887, which used the term "agricultural science". The Hatch Act was driven by farmers' interest in knowing the constituents of early artificial fertilizer. Later on, the Smith-Hughes Act of 1917 shifted agricultural education back to its vocational roots, but the scientific foundation had been built. After 1906, public expenditures on agricultural research in the US exceeded private expenditures for the next 44 years.

Genetics

A genetic study of Agricultural science began with Gregor Mendel's work. Using statistical methods, Mendel developed the model of Mendelian inheritance which accurately described the inheritance of dominant and recessive genes. His results were controversial at the time and were not widely accepted.

In 1900, Hugo de Vries published his findings after rediscovering Mendel's work, and in 1905 William Bateson coined the term "genetics" in a letter to Adam Sedgwick. The study of genetics carried into an experiment isolating DNA.

Contemporary

Agronomy and the related disciplines of agricultural science today are very different from what they were before about 1950. Intensification of agriculture since the 1960s in developed and developing countries, often referred to as the Green Revolution, was closely tied to progress made in selecting and improving crops and animals for high productivity, as well as to developing additional inputs such as artificial fertilizers and phytosanitary products.

However, environmental damage due to intensive agriculture, industrial development, and population growth have raised many questions among agronomists and have led to the development and emergence of new fields (e.g., integrated pest management, waste treatment technologies, landscape architecture, genomics).

New technologies, such as biotechnology and computer science (for data processing and storage), and technological advances have made it possible to develop new research fields, including genetic engineering, improved statistical analysis, and precision farming.

Education

Italian Renaissance book on cultivation, written in 1549

There are various universities around the United States which are well known for their education of the agricultural sciences. These universities include Texas A&M, Stephen F. Austin State University, University of Idaho and many others.

References

- John Armstrong, Jesse Buel. A Treatise on Agriculture, The Present Condition of the Art Abroad and at Home, and the Theory and Practice of Husbandry. To which is Added, a Dissertation on the Kitchen and Fruit Garden. 1840. p. 45.

- Hillison J. (1996). The Origins of Agriscience: Or Where Did All That Scientific Agriculture Come From?. Journal of Agricultural Education.

Permissions

Index

www.ingramcontent.com/pod-product-compliance
Lightning Source LLC
Jackson TN
JSHW051815140125
77033JS00048B/225

* 9 7 8 1 6 3 5 4 9 0 1 8 3 *